Dermatological Atlas of Indigenous People

Marcos Cesar Florian • Jane Tomimori
Sofia Beatriz Machado de Mendonça
Douglas Antonio Rodrigues

Dermatological Atlas of Indigenous People

Springer

Marcos Cesar Florian
Universidade Federal de São Paulo
São Paulo, Brazil

Jane Tomimori
Universidade Federal de São Paulo
São Paulo, Brazil

Sofia Beatriz Machado de Mendonça
Universidade Federal de São Paulo
São Paulo, Brazil

Douglas Antonio Rodrigues
Universidade Federal de São Paulo
São Paulo, Brazil

ISBN 978-3-319-59444-6 ISBN 978-3-319-59446-0 (eBook)
DOI 10.1007/978-3-319-59446-0

Library of Congress Control Number: 2017943507

Printed on acid-free paper

This Springer imprint is published by Springer Nature
The registered company is Springer International Publishing AG
The registered company address is: Gewerbestrasse 11, 6330 Cham, Switzerland

Preface

The Escola Paulista de Medicina/Universidade Federal de São Paulo (EPM/UNIFESP), Brazil, has developed health assistance activities and teaching and research programs among indigenous people since 1965. It is a Brazilian university with considerable experience in that area.

The Brazil Unified Health System (SUS) calls for universal care mainly in primary care. The health of indigenous people has received special attention from the Ministry of Health, and special Indigenous Sanitary Districts have been established throughout the territory of Brazil.

Professionals who work in the field of indigenous health should see that this subsystem works and is viable. The Xingu Project from EPM/UNIFESP has worked intensively on raising awareness of and training professionals to work on the health of indigenous people. The extension courses for interested health students are a part of this work.

Proposals regarding knowledge of and intervention in health problems affecting indigenous communities are ambitious. They involve holding a continuous dialogue with different cultures in order to understand their worldviews, their healing systems, and their conceptions of health and disease. In addition, we need to develop actions in primary care to minimize the negative health impact arising from contact of indigenous peoples with the broader society and to adapt health intervention models in application to different indigenous groups and their social realities and lifestyles. Brazil is home to 230 indigenous peoples speaking 180 languages, a group of about 600,000 people. This is the main action field of the Xingu Project.

Among the various activities related to the overall health of these people, the experience and photographic documentation recorded over the years have allowed the authors and contributors to gather rich materials related to some skin diseases that affect these populations. This *Dermatological Atlas of Indigenous People* is a result of the work conducted over several years. We have as our goal the production of material that sheds light on the clinical-dermatological aspects and the culture of these peoples. The therapeutic management of various clinical situations described here has always strived to incorporate the best and latest clinical practices, but whenever possible taking into account their economic viability because many indigenous populations live in less-developed countries.

This edition of the *Atlas* benefited from the invaluable participation of the following collaborators:

- Roberto G. Baruzzi (in memoriam) MD, PhD. Full Professor, Department of Preventive Medicine and scientific adviser to the Xingu Project, EPM/UNIFESP, Brazil.
- Helio C. de Mello. Photographer and collaborator, Xingu Project, EPM/UNIFESP, Brazil.
- Mariana D. Batista MD, PhD. Dermatologist, Department of Dermatology, EPM/UNIFESP, Brazil.
- Solange M. Maeda MD, M.Sc. Collaborator dermatologist, Department of Dermatology, EPM/UNIFESP, Brazil.

São Paulo, Brazil

Marcos Cesar Florian
Jane Tomimori
Sofia Beatriz Machado de Mendonça
Douglas Antonio Rodrigues

Acknowledgments

- Arnaldo L. Colombo MD, PhD. Full Professor, Department of Medicine– Infectious Diseases, EPM/UNIFESP, Brazil.
- Dalva Regina Neto Pimentel MD, PhD. Professor, Department of Dermatoloy, EPM/UNIFESP, Brazil.
- Mílvia M.S.S. Enokihara MD, PhD. Department of Pathology and Department of Dermatology, EPM/UNIFESP, Brazil.
- Olga F. Gompertz, PhD. Professor, Department of Microbiology, Immunology and Parasitology, EPM/UNIFESP, Brazil.
- Orion Sant Ana Motter Borba. Mycologist, Department of Microbiology, Immunology and Parasitology, EPM/UNIFESP, Brazil.
- All professionals of the EPM/UNIFESP who work with indigenous peoples.
- All of the indigenous nursing workers and health indigenous agents who work in the Xingu Indigenous Park, Brazil.
- All of the indigenous peoples.
- CASAI (Casa de Saúde Indígena) from Juara City (Mato Grosso State, Brazil).
- FAPESP (Fundação de Amparo à Pesquisa do Estado de São Paulo) by the support for this project (Process 2006/03908-0)

Contents

Author Biographies

Marcos Cesar Florian MD, Dermatologist, Leprologist, MSc, PhD.
Affiliate professor, Department of Dermatology, EPM/UNIFESP, São Paulo, Brazil.

Coordinator and Professor, Faculty of Medicine, Universidade Nove de Julho, São Paulo, Brazil.

Jane Tomimori MD, Dermatologist, Leprologist, MSc, PhD, PostDoc.
Full Professor, Department of Dermatology, EPM/UNIFESP, São Paulo, Brazil.

Assistente étrangère, Université Paris VII, France.

Sofia Beatriz Machado de Mendonça MD, Public Health Physician, MSc.
Coordinator, Xingu Project, Department of Preventive Medicine, EPM/UNIFESP, São Paulo, Brazil.

Douglas Antonio Rodrigues MD, Public Health Physician, PhD.
Coordinator, Health and Environment Unit Department of Preventive Medicine, EPM/UNIFESP, São Paulo, Brazil.

1 Social Representations of the Body, the Skin, and Related Diseases in Indigenous Thought

Thinking of the body, you can mistakenly face it as purely biological, a universal patrimony whereupon the culture would write different histories. After all, men of different nationalities show physical similarities. However, in addition to physical similarities or differences, a set of meanings that each society writes on their members' body through time, these meanings define the body in various ways. Daolio, 1995

The original idea of this book arose from the observation of the high prevalence of skin diseases in the daily routine of health professionals who work for the Primary Health Care in Amazon's indigenous areas and the absence of adequate support material for these professionals.

As a result of an action-research called "Dermatologic Investigation and Evaluation of Jorge Lobo's Disease Evolution in Caiabi Indigenous People, Central Brazil," financed by the São Paulo Research Foundation – Fapesp (process 2006/03908-0), which had as its objectives, besides reviewing the Jorge Lobo's disease cases, the performance of an investigation into the prevalence of dermatosis in the villages of the middle, lower, and upper regions of the Xingu Indigenous Park and the local health staff's capacity to manage mainly skin diseases found in their daily local services. The production of this atlas was made by the cooperation of the Xingu Project, the Department of Preventive Medicine, and the Department of Dermatology of Escola Paulista de Medicina/Unifesp (EPM/UNIFESP).

It seemed important to present in this book, besides scientific knowledge, the indigenous people's vision of their own bodies and skin, the way they think, classify, and interfere with the health-disease process, and which therapeutic itinerary they adopt to treat skin-related issues.

1.1 The Body and the Skin: Center of Social Signs

The body and, in particular, the skin are a person's introduction to their environment and community, which is why it is believed to represent an individual's contact with the external, ruled by culture. Such signs as tattoos, paintings, and scarifications can be registered on the body and reveal what group a person belongs to, her social status, or even her state of mind.

As a canvas, the skin can display several drawings, combining colors and adornments, depending on the moment, one's state of mind or health, or the ritual one is participating in. In general, during mourning periods the body is not painted or adorned, thereby revealing sorrowful feelings.

According to David Le Breton , in his book *A Sociologia do Corpo*[1] on bodily inscriptions:

> Social and cultural bodily markings can be completed by direct writing by the collective on the actor's skin. It can be made in the form of removal, deformation, or addition. This symbolic modeling is relatively frequent in human societies: ritual ablation of a bodily fragment (prepuce, clitoris, teeth, fingers, tonsure, etc.); epidermis marking (scarifications, incisions, apparent scar, infibulation, teeth modeling, etc.); tegumentary inscriptions in the form of permanent or temporary tattoos, make-up, etc.; body form modifications […]; use of jewelry or objects that deform the body: reed rings and pearls that cause, to the individual growth, a neck extension, inserting disks in the upper and lower lips […]. These bodily markings fulfill different functions in each society. Seduction instruments are more often a ritual way of indicating affiliation or segregation. They symbolically integrate people within their community or clan, separating them from other communities or clans and, at the same time, the surrounding nature. They humanize the person, putting him socially in the world […]. They reproduce the social status or, more specifically, the matrimonial status in a world that can be read by everybody.

Regarding the body and the skin as the center of social meanings, Clastres[2] reveals another perspective, the body "as written surface […] able to receive the legible law text,", obedience to rules and social laws, particularly when referring to the role of suffering and scars. The body is the memory, for example, in the initiatory rituals of indigenous societies, scarifications, tattoos, or other procedures that leave a mark, a mark that tells the story of the subject's life. Changes in the body and skin occur during every biological

[1]Le Breton [13].

[2]Clastres [14].

M.C. Florian et al., *Dermatological Atlas of Indigenous People*, DOI 10.1007/978-3-319-59446-0_1

cycle, and such changes are for adolescents the most important moments for modeling purposes.

> In the initiatory ritual, the society prints its mark on the youngest bodies… The mark is an obstacle for forgetfulness, the body brings printed on itself the skin grooves memory – the body is a memory. The mark proclaims surely its belonging to the group… The society dictates its laws to the members, inscribes the law texts on body surface […]. The mark on the body, like on all bodies, states: "you will have no desire for power, neither covet be submissive", and this law does not separate and can be only inserted in a non-separated space: the body itself (Clastres, 2000, p. 196).

Among the *Ikpeng*, for example, children around seven and nine years old must have a tattoo on their face, which is a characteristic mark of this people, in a rite of passage. The whole village mobilizes for the party:

> The main party celebrated by this people, the *Ikpeng*, is *Moyngo*, in which boys' faces are tattooed. The ritual is preceded by many dance sessions and, in the end, by a big hunt, in which the to-be-tattooed children's fathers, who are the hosts, participate. After about 1 month, an expedition messenger is sent to the village announcing the hunters' return. The next day, during a dance session, to the sound of flutes and the indigenous chief's voice, the hunters arrive with a huge basket full of prey.
> The hunters camp next to the village and women go there to take hunt made in a *moquém* and leave *beijus*. The participants cover the body with some wood resin and stick bird feathers on it. They go in the village at nightfall and drink sweet *perereba* (kind of porridge). Next, each man dances holding in one hand a to-be-tattooed child and in the other a torch. Again, they spend a whole night dancing. Lastly, at the end of the party in the morning, the children are tattooed. First, they make incisions (stripes) on the children's faces with a *tucum* thorn and then they pass the coal extracted from the courbaril resin (*Instituto Socioambiental*, undated).

1.2 The Skin as the Body Limit and the Exchange of Substances

Concern for the body, and the skin in particular, is huge among indigenous peoples, aesthetically and functionally. The aesthetics and beauty are adored from the beginning of life, in early childhood.

The skin, the body's boundary and place of contact with the external, also has a key role in the protection of the body and, at the same time, works as an entry way for medicines. Many procedures are performed on the skin to protect the body using special paints, mainly on children, using *pequi*, *tucum*, or maripa palm fruit oils. Annatto or saffron, for example, keeps bad spirits and insects away. Annatto paste, genipap, bird feathers, certain tree resins, and cotton are used as clothes or costumes in rites and parties. For indigenous people, a painted body is a healthy body.

Body modeling starts in toddlers, tying the leg under the knee, so that the child's leg will get more beautiful. Tying is also done in the arm, at the height of the deltoid muscle insertion, so it becomes more beautiful and rounded.

For numerous disease treatments, the skin constitutes a particular part, either for removing inoculated objects by *mama'e*or spirits or for absorbing the healing characteristics of plants. In many therapeutic procedures, baths are performed with herbs, leaves, tree bark, or roots. The skin can also be scarified to purify the body by removing dark blood and what it contains; to eliminate laziness, one of the factors that cause diseases; and to enable the entrance of medicines.

Next, the care provided for body and skin used in different events and stages of community life are discussed.

1.3 The Process Health-Disease and the Social Representation of Skin Diseases

During the process of organizing and implementing health services in indigenous areas, starting with scientific knowledge and biomedicine, quite different healing systems and therapeutic itineraries were found.

The vision of the progression of an illness and the conception of the body and health these peoples have relate to their world vision. Disease does not exist outside of the sociocultural context, health-disease interplay happens from the representation of disease in each society and reveals itself, basically, in three dimensions: subjective, biophysical, and sociocultural. These dimensions consider the feelings of being sick, the signals and physical symptoms, and the sociocultural group perceptions of being sick. These three realities interact continually in a dynamic and procedural form [1].

To the indigenous people of Xingu skin diseases are associated with nourishment, violations of social rules, contagion from other sick people, and spells, depending on the kind of wound and its evolution, according to interviews with health officials together with wise men, shamans and herb doctors, from their communities. More recently, they have started to associate some kinds of skin disease with a lack of self-hygiene, mainly related to the use of clothes, blankets, and hammocks, without the necessary care.

In many stories, the etiology of a skin disease is related to animal spirits and other nature spirits that frequently have something in common with the wound. For instance, when eating spotted eagle ray, it is common to see round and whitish wounds on the body, just like the ray. The same thing happens with the lowland paca, which has "spots" all over its body. In another example, if a pregnant mother eats honey from a certain kind of bee, her child will have wounds on its body resembling the bites of this kind of bee.

Besides physical similarities to nature spirits, the explanations given for the appearance of skin diseases lead us to origin myths, which exist in different versions depending on the storyteller's knowledge. At any rate, they reveal and give meaning to many rules of behavior and relations to other people, and transgression of those rules causes disease.

These stories are transcriptions from interviews with students of the course of indigenous health agents from the Xingu Indigenous Park, as field research work in their communities, on the topic of skin diseases, intestinal parasites, and environment, conducted in June 2007, at the Pavuru plant of DSEI Xingu/MT.

One of the explanations for a skin disease called *mirukai* in the Kaiabi people's language burn similar to erysipelas, has its explanation in the myth of food origin, as the health agents tell us, based on their interviews with the oldest and wisest men from the community:

In the past, there were no foods like corn, cassava, peanut, fava beans, pepper, potato, yam, mangarito, sweet cassava, cotton, calabash and *janyrũ*, which is a kind of calabash used to keep tucum oil. The Kaiabi people spent many years without these foods. People lived starving. They only ate wild fruits, for example, maripa palm fruit, tucum, moriche palm fruit, chestnut, cocoa, *banana brava* fruit, and honey.
Many years passed until the children of a widowed woman, named Kupeirup, as they went out to the bushes looking for honey and fruits, could not find anything else; it was hard to find something, because the fruits they had planted did not grow fast. For this reason, they did not have any fruit to eat in that region, and their mother, worried, said:

"Boys, I am very sorry for you, because you are not finding fruits and honey. The trees you planted will not grow soon, they will take some time to bear fruit. Therefore, I ask you to prepare a big plantation field. When you set fire, take me to the middle of the plantation field and burn me as well. When my body burns up, it will grow farm products for you."

They became sad about what their mother told them and asked if it was necessary:

"Why are we doing it to you, mistreating you? You are our mother. We just want to take care of you. We are not doing that to you."

Their mother repeated:

"If you do not do that to me, you will never have food."

They answered:

"We will miss the food, but we cannot burn you up. It will not be good for us."

Their mother repeated:

"My sons, I know you are worried about me, but you need not worry. I ask you to burn me, because I will return to you."

The sons said:

"We will do it, mother."

However, they became sad because of this decision.

They made a big plantation field. First, they cleared the land, then took it over. After overthrowing the owners of the plantation field down, they told their mother. She repeated:

"Boys, it is time to burn the plantation field, then you will burn me. Do not be sad because of me, I will return to you. I am going to become a lowland paca and I will be around the field. At that time you have to set a trap and catch me in the trap. Do not let me escape. If this happens, I will not return to you. When you burn the plantation field, scream to burn me. When I hear your shouting, I will sow the plantation field. After burning the field, you must go far from there, choose another place to live and grow fruit to eat."

That is how she instructed her sons.
She continued explaining what they were supposed to do:

"Time will pass and you will see a caica parrot flying over your heads. This will be the first alert, which means food is growing. Then, two caica parrots will fly over your heads, telling you about some crops, like corn, that are growing. After some time, many caica parrots will fly over your heads. This will be the signal that the crops will be ready for you. Therefore, from that day you will not starve anymore, because many crops will grow."

Kupeirup explained to her sons about food preparation and taught them how to conserve seeds and seedlings, so they would not consume them all, keeping the crops, so that when they cleared the land and planting time came, the seeds would be well preserved. She also taught her sons to take good care of the crops so they would never be consumed or disappear.

She repeated many times her instruction to her sons. Once she finished instructing them, she said:

"The time to burn the plantation field has arrived. Now you can take my hammock and hang it in the middle of the field, I will be lying there."

They took their mother's hammock and set it in the middle of the field. After lying on the hammock, she asked her sons to set fire to the field. As the field was burning, they screamed and called their mother's name, saying:

"We are burning the field for you, mother, lady Kupeirup Kooooo!!!"

The boys were on the edge of the field watching the fire, and when it got to Lady Kupeirup, they heard an explosion, which was the food spreading out over the field as if in a tornado.

When the noise died down, they turned away from the field, heading to the camp where their mother asked them to stay. The place had a great abundance of wild fruits. They stayed there, passing the time and following their mother's advice.

The time passed, until a caica parrot flew over their heads. So they became happy because they knew the crops were growing. After some time, two caica parrots flew around. They said:

"Our crops are almost ready, the corn is about to blossom, and the fava beans may be green beans."

They kept on talking about the plantation field. After 1 month, many caica parrots started flying over their heads. Then they went to check the field. When they arrived, they saw all the crops and said:

"Our mother planted the crops for us! Now we can look for her."

They started looking for their mother around the plantation field and found her in a hole in the ground, transformed into lowland paca. Then they started to prepare a trap to catch her. When they finished, they put the trap in the hole to catch their mother. Suddenly, the *Maramu'jangat* appeared, the Changer. At that time there were not many kinds of animals, and this entity transformed people into animals, so that there were more animals. The Changer asked the boys:

"What are you doing?"
"We are going to catch our mother."
"Let me catch her for you," said the Changer.
"No, we will catch our mother ourselves."

The *Maramu'jangat* pretended to leave, but in fact, he was hidden. When they set the trap and tried to catch the lowland paca, the Changer came up and scared them:

"What are you doing?"

That is how they let their mother escape, and she ran away, screaming like a lowland paca:

"Wee, wee, wee, wee."

The *Maramu'jangat* changed lady Kupeirup into a paca. He said to the boys:

"The lowland paca will always want to be close to the plantation field and eat a lot of corn."

That is why until now lowland pacas enjoy being near old plantation fields and eating farm corn. In the past, boys and girls did not used to eat pacas, only elderly people did, but nowadays, young people eat it. However, the lowland paca can cause burns on people's legs, because it was burned in a plantation field. Thus, because of Lady Kupeirup, the crops began to grow:

- one of her teth became corn,
- her hair became corn hair and cotton,
- her fingers became peanuts and pepper,
- her legs became cassava,
- her hands became cassava leaves,
- her head became calabash,
- her brains became *cará*,
- her thighs became sweet cassava,
- her milk became sweet cassava water,
- her foot became elephant ear fruit,
- her tongue became fava beans,
- her heart became potatoes.

In this way, Lady Kupeirup made crops for the ancients and made the crops we know to this day. From this woman came the myth that the paca is not good food; it causes harm like *mirukai* on people's bodies. Therefore, the *mirukai* catches people only when they eat lowland paca (Aramut, Arasi, Aruta, Kwariup, Pirapy, Tafarejup, Tamakari, Wyra'wat e Yefuka Kaiabi, 2007).

Origin myths serve as explanations of numerous social rules on behaviors, attitudes, relationships with others, and nourishment and often explain prohibitions and special diets that must be followed at each life phase to guarantee good health. Violations of these rules generally cause disease.

According to some studies conducted by health indigenous agents, many diseases are associated with forbidden foods. Some diseases, particularly the most complex and worst ones, are associated with spells. What follow are some explanations for the appearance of skin diseases.

The Kaiabi People, according to Aramut Kaiabi:

According to my people, skin disease appears in any time of someone's life because people do not follow some rules older people impose on younger people. The rules are related to some foods that cause skin disease on people over their lives, whether child, adult, or elderly. I will give an example: according to my society, eating lowland paca causes skin disease, which is called *aekai* (or *mirukai*) – burn – which can affect any part of the body. The skin disease also occurs on newborns; a pregnant woman eats honey from *mamairowasing*, *tataeit* bees, and other species. This honey can cause wounds on children's bodies after birth. When an older person or a child mixes some foods, for instance, when they eat monkey meat with agouti meat, this can also cause skin disease, *pitemunat*. My people knew these diseases. With time, the *miraip* and the *tyra* appeared among the Kaiabi. *Tyra* is a disease that seems like mycosis, hypochromic, for which there is no cure and remains on the body until death. *Miraip* is an infected wound over the entire body. This disease appeared because there were too many fights between the Kaiabi people and other indigenous people, in which were killed many people. In this way, one day, they killed a person who had this disease and, as the Kaiabi liked to celebrate when they warred, they brought the body to the village to celebrate the enemy's death. During these parties, they composed songs in verse. In the lyrics, they told the story of the war between the Kaiabi and other peoples. This disease, *miraip*, was brought to the Kaiabi people through the enemy. These two diseases, *miraip* and *tyra,* were incurable. People who had these diseases tried to cure them many times, but unfortunately, no wild medicine could cure them. However, after contact with white people, the diseases declined because they found that it was Jorge Lobo's disease, which they had some treatment for. Other skin diseases are frequent nowadays because people do not wash their clothes, for example. They obtain clothes from other people and do not wash them, either before or after wearing them. Or when a person has a skin disease and another lies on that person's hammock, he will also catch the disease.

The Kamaiurá People, according to Sula Kamaiurá:

The skin disease happens when the wizard makes spells on a person and makes people allergic to some kinds of food, like ray, *jaú* fish, honey, and others. The honey can cause wounds and make children weepy. It happens when mothers and fathers do not respect rules about eating related to pregnancy.

The Kisêdjê People, according to Poan Trumai Kaiabi:

Skin disease according to the Kisêdjê people occurs because people do not respect rules and do not take care of themselves (lack of hygiene). For each kind of disease, there is a specific history.

The Yudjá People, according to Tawaiku Juruna:

According to the Yudjá people, some kinds of skin disease are transmitted through nature. These diseases appear on people according to the environment and the places they go. They affect people who hunt or those who have contact with various types of plants that can cause allergies and cause wounds on the skin. Some plants cause itches, which disappear by themselves. Other skin diseases are caused by food; for example, the *jaú* fish causes on children under two years old one kind of skin disease, which often happens when the parents do not follow their people's traditional rules. This kind of wound may happen on the groin or the armpit region, and it is very irritating. Butterflies can also cause skin disease. Children or adults who touch the wings of some butterfly species and then touch their bodies with unwashed hands can be affected by white spots on their skin, over their whole body. The way an owl cleans itself can cause mycosis on people, because they are not paying attention and walk where the owl was. Some skin diseases also appear because of dirt in a river. There is a skin disease that appears on toes, which for my people is caused by rainwater. Papaya, *cará*, and squash are forbidden foods for small children, because they can cause skin problems.

The Trumai People, according to Mataripé Trumai:

For my people, skin diseases appear because of the following problems: eating food the body does not accept or because the person has never eaten these kinds of food, for instance, red spotted eagle ray; when the parents do not follow traditional rules on pregnancy and breastfeeding; when a person takes a bath in a dirty or polluted river. The ancient Trumai people believed that remain sat at their houses' doors, the risk of catching *tanakura*, which is a chronic wound, increased. If treatment with medicinal herbs was not carried out, the disease could become worse and harm the person's health. The ancient people also believed that dreaming every day about scratching oneself with *peixe-cachorro (arranhadeira)* teeth was a signal one would catch *yawhuthut*, a reddish wound on the skin that gets worse with time, when it gets white and at the same time flakes off from scratching.

The Wauja People, according to Apayatu Wauja:

Among my people, a dangerous skin disease appears when a person puts *kuretse* (spell) on copaiba oil, which is used in painting parties.

The Ikpeng People, according to Mekirinpô Ikpeng, Piwara Horotowi Ikpeng, Melo-bo Kamaritá Ikpeng, and Davids Penewô Y. Txicão:

According to my people, skin disease comes into existence with its appearance. Skin diseases do not appear by themselves on people; they affect people who do not respect traditional rules or who eat *recongo* meat, Kulu ant eggs, parts of birds' backsides, wasps, lowland paca, or capybara. Skin disease can exist since birth, depending on the foods parents eat. When parents eat fish called *yetepun*, *pitpirak*, and *tangren*, children can become allergic. Skin disease is also transmitted when lying on dirty, contaminated hammocks. That is why we do not lie down on other people's hammocks.

1.4 Therapeutic Itinerary

The therapeutic procedures for skin diseases are used according to wound type, number of skin lesions on the body, risk, and evolution. Prevention and treatment start at home, in the family, using herbs, roots, oils, and substances of common knowledge, for instance, saffron, annatto, *tucum*, and copaiba. During the process of disease evolution, other people can be consulted, for example, herb doctors or owners of a specific herb or medicine who hold some specialized knowledge and know how to ask spirits or the owners of medicines permission to use their medicine and cure their disease. Supplicants know the right prayers for any type of problem, like the prayer to heal a boil among the Kisêdjê people. Shamans (*pajés*) are called when a disease requires intervention from spirits and communication to the supernatural or the invisible world, according to Carmen Junqueira [2]. Nowadays, Western health professionals, for example, health agents, nursing assistants, nurses, or doctors, are called to treat skin problems and many times they work with traditional masters. We call the approach taken by the patient or their relatives who seek healing "therapeutic itinerary", as discussed by Langdon [8].

What follows are some stories about skin disease classification and their treatment.[3]

The Kisêdjê People, according to Poan Trumai Kaiabi:

According to the Kisêdjê people the skin diseases are as follows:

Wakrâyaká: white spots, similar to mycosis wounds. To the Kisêdjê, these spots, which appear on people's bodies, are caused by a sick spotted eagle ray; for this reason the Kisêdjê do not eat this type of ray; they only eat the normal ray, which does not have (white) spots. The treatment for this disease consists in heating a stick, which is used for making an arrowhead called *hwadô*; a newly cut one is commonly used because the one used in the arrow has no more liquid; it is already dry. After the stick is cut, it is placed on fire coals to extract the liquid it contains. When the water is almost completely extracted, they place the stick on the spots. They continue with this procedure until the spots disappear.

Wakrajkasaktxira: mycosis. To the Kisêdjê what causes this skin disease is the jaú fish, which is why we do not eat it. A root called *kukhãj nhimonsõ* treats this mycosis; it must be heated in fire, softened while still hot, and put on the mycosis. The same herb can also be used to treat chilblains on the feet, *wahwaj tomtomtxi*.

Wanhikotxi: abscess on skin. To the Kisêdjê, according to the shamans' wisdom (*mêkatwân khetdjê*), this skin disease is a plague from a nature spirit, which goes house by house searching for lazy people and those who do not wake up early. It is an infection that comes from inside the skin, accompanied by swelling, pain, and fever. When the time is right, a sharp-pointed *horé* (arrow) must be used to stick the

[3]These stories were extracted from the Field Researches performed by the Indigenous Health Agents (IHA), who are part of the graduation course developed by Unifesp at the Xingu Lower, Middle and Upper coverage area, DSEI Xingu/MT.

wound. After removing all the pus, a moriche palm fruit thread is applied, so it does not heal completely, facilitating pus exit. When the person realizes the wound is better, the moriche palm fruit thread is taken out, so it can heal. It is equally important that a prayer be said by a supplicant to reduce the edema. Potato and cassava leaves are also used; they must be heated in fire and applied to the wound, so that the edema and the infection do not increase.

Kusẽ: furuncle (boil). This is an infection on the skin. It heals from the use of procedures similar to those performed on *wanhikotxi*, unless one is using a *horé* to stick it. According to older people, eating too many eggs facilitates the appearance of a furuncle.

Mbeni: a kind of infection in the subcutaneous skin, which causes edema, hyperemia, high fever, pain, and, after some time, a wound around it with many spots, which are typical of this disease. A person who contracts this disease cannot eat any kind of sweet food, for example, bananas, potatoes, or papayas. According to one myth, if a person eats any kind of sweet food, the disease will progress very fast and may cause death. This disease is not transmissible, but as soon as it is contracted, it must be treated by professionals capable of diagnosing it and have neither fear nor pity. It is not easily diagnosed; sometimes it looks like a common wound and is treated with medicinal plants, but the wound gets worse. When detected at an early stage, someone specialized in cutting search for shells found on river bottoms can, after finding a shell, remove the wound from the patient by cutting the wound deeply with the shell. The patient feels a lot of pain, because there is no anesthetic. This is the only treatment, and if the injury is not completely healed, the disease returns. When it is too advanced, it cannot be treated, and may cause death. To the Kisêdjê, what causes the disease is a kind of greenish jam that is under water, sometimes stuck to a rock, and can seem like honey wax, which is why it is called *mbeni*. Only older people know how to identify the disease. This is why people cannot take this jam. Parents instruct their children not to play with this jam. Many times they play with *mbeni if* they find it in a river, but the disease can take years to emerge. No medicinal plants are used in the treatment of this disease.

KrâKabentxi: a skin inflammation characterized by many edemas, pain, hyperemia, and fever on the wound site. When not treated, it can become infected, causing an ugly scar (somewhat deep). It can be treated with *tucum*. First, *tucum* is burned in fire and the person must expose the wound to the smoke, until the *tucum* becomes coal. Then the *tucum* coal is mixed with *tucum* oil. This mixture must be put on the wound area until it gets better. When the person who is treating the inflammation is a supplicant, he applies the mixture and prays, so that the wound heals faster.

Kupekratxi: any injury that becomes a wound can become infected, with or without pus stocks. It can be caused by a mosquito bite or lice, which can cause infected wounds on the head. The sap of a tree leaf called *hwisôsâyahoró*, which is extracted when pulling the leaf from a branch, can be used to treat it.

Mênhôtôtxi: moniliasis. These are small white plaques similar to milk stuck in the oral region. This disease is treated by fluid from a banana or *hwĩkajkhirê* – a kind of softwood – that is cooked and put in the patient's mouth.

The Kaiabi People:

The skin diseases my people know are furuncles, mycosis, escabious, *miraip*, *pitemunat*, *ae jap*, *ae kai,* measles, and chicken pox. In the past, my people did not know measles and chicken pox. Therefore, they do not know how to treat them. There is a traditional treatment for each disease, and there is no treatment for some. Furuncles can be treated with *muanpasing* and *ysineyp*; the root must be peeled and put in place, in which is stuck so that the pus comes out. The root must be placed on the wound until it gets better.

There is no traditional treatment for *miraip*, but it can be relieved or controlled. A piranha tooth is burned and then some water is put on the ashes. Then this water with the ashes must be placed on the wound.

The treatment for mycosis, *tyra*, is *pyaupiat* – a bitter greenish liquid, that is stuck to fish liver. The wound must be scraped before application of the medicine.

The traditional treatment for *ae jap* – wound – is burning a cotton leaf and seed and *iniaywet.* The wound is placed over the smoke. The smoke enters the wound, helping it to heal. For *pitemunat,* the traditional treatment involves taking a type of fish stupefying plant, squeezing out its poison, and bathing the patient in this liquid. The *aka'a kasing* leaf is also used: besides bathing with it, the shaman also puts it on the patient's body, praying and talking to the spirits.

The traditional treatment for *ae kai* or *mirukai* – burn – is to take a gourd, cutting it up, and removing the core, which must be put on the wound site. Another treatment for this disease is hunting, killing, and extracting the blood of a *jacutinga* and putting it on the wound.

The Yudjá People:

Skin diseases include *iya-ibidamã, nasusu-ibidamã, pitxa- urahïhï-ibidamã, pewapewa, bi'ata'ata, wareware-seha, pïata,* and *kïrïta* and *huï'ï-apeta*.

Iya-ibidamã is normally treated with maripa palm fruit oil mixed with annatto. To cure the patient, use a climbing plant called *iyanauka* that grows near rivers. *Bi'ata'ata* is the name of the itch, prurigo. *Pïata* is treated with a medicinal herb, mainly a small fruit. *Kïrïta* treatment is made with small-fish feces (*carazinho*). *Pitxa-urahïh* is treated with saffron and a type of ginger. *Pewapewa* is treated with a kind of roasted potato (made in a *moquém*); another way to cure this disease is by touching the patient without warning her; if she gets scared and starts to scream, after a few days, the disease will disappear.

The Trumai People, according to Aramy Trumai Kaiabi:

There are many skin diseases in Trumai knowledge. My mother told me that *yawkitsikitsin* is caused by a mosquito bite (*muriçoca*). Common wounds, *yawelemxo*, are caused by ant, tick, and mosquito bites. *Yawhuts* is a kind of itch that makes the skin dry, thick, and peeled – what it looks like my mother did not tell me. *Yawpatsae* are whitish and reddish spots that appear on the body when someone eats foods that are bad for the skin. An abscess (*kutsut*) appears on the body like ant bite, and it is painful. These are the skin diseases in Trumai knowledge.

There are medicines for all skin diseases known to the Trumai people: for *yawhut*, we use a root called *saruiawiwal*. This root is peeled, mixed with annatto or *murukuiu* and water, and placed on the wound hot, but first the skin must be scratched, so that the root liquid can enter the wound. During treatment, certain fish, for example *jaraqui, curimba*, and *piau*, cannot be eaten. We use the root until the itch goes away.

For white spots (*yawkïtsikïtsin*), there is another treatment we call *tsinapai*, which is a tree near a river. From this tree we take a paste, which must be mixed with annatto and water, and applied to the wound heated. There is also a root called *tsinonmaskakwach*, a kind of passion fruit found in the jungle, that is also used for these diseases.

For *tanakura* there are two medicines: the first, called *tanakuraoken*, is in a calabash. It must be opened when the sun is hot; the seeds are removed, smashed, and mixed with a little water. The mixture must be warm when applied to the wound. Another medicine is called *tehneneiat*; it is similar to *jocote*, which is also used in the treatment of this disease.

Abscess treatment (*kutskut*) is simple. We use the cotton seed and a root. To put it on the wound, the seed must be thoroughly softened and mixed with water. The medicine must be hot. A moriche palm fruit thread is also used and left in the wound, but first a hole must be made in the wound using an *arranhadeira* tooth. Then the thread is inserted, so it is easier to remove the pus.

For wounds (*yaw elemxo*), treatment is made from a small fruit that grows near rivers. The fruit is taken, peeled, smashed, and mixed with water, like other medicines. The mixture must be put on the wound hot.

The treatment for wounds with whitish and reddish spots (*yawpatsae*) consists of two roots; the first is *ika'dapatak*, large leaf, while the other is *ika'dasupi*. These two roots are used in the same way as other Trumai traditional treatments.

The Kamaiurá People, according to Sula Kamaiurá:

There are many types of skin diseases, but people do not know the proper name in the Portuguese language. In the indigenous language they are *tsitsinga ruijap, miruru* (wound), *miremona ruijap*.

There are many types of roots for these diseases; it depends on the herb doctor diagnosis. We use roots: *janyparan*, *yrykuran*, *yrykujup*, and *ka'api'i*.

To Ikpeng People, according toMekirinpô Ikpeng:

The skin diseases in the knowledge of my people are *pïlawat*, a very dangerous disease called leishmaniasis; *antango* is also a dangerous disease and can cause death; *peri* is a common wound; *turi* is a chronic disease, it never heals, it dries for a few days, but it returns; *mutu* is a disease on the "external body", it can itch and it hurts a lot; *orek* is a hurting wound, without pus, inside the skin, it can swell a lot and get bigger.

Nowadays, there are some skin diseases and others no longer exist.

The traditional treatment for my people is using jungle herbs, for example, *kurium, tarik, angejewï*, and *walap*. These are medicines for skin diseases my people know.

1.5 Final Considerations

This reflection seeks to provide subsidies for fieldwork and indicate ways to address problems related to skin diseases among indigenous people.

> An interesting event I experienced involved a child with an extensive burn on his arm who went to a Basic Health Unit (BHU). The family had used something they did not recognize, I decided to clean the arm and bandage it; the next day the child returned to the BHU without the bandage and again with the same substance. I talked to the mother about the importance of keeping the wound clean and I put on another bandage. They returned and the situation was the same as before. Then I talked to the mother together with an Indigenous Health Agent (IHA) about what they were taking, and we decided to continue with the traditional treatment. We instructed the family to keep the wound covered, and the IHA would observe the inflammation. In a week, it was completely healed. I only understood what was happening when I asked what medication they were applying to the burn, and they explained me the procedure, what was taken, and what it was for. After that, I stopped imposing on them what I thought was right and respected their culture and reconciled my cares with the mother's [story of a local health team nurse].

For the attendance of the cases and therapeutic interventions, it is essential to understand how indigenous people interpret the world and their traditional therapeutic procedures. Dialogue is always important so everyone can agree on a course of action to reach a positive outcome.

Comprehension of the health-disease process in its different dimensions, the associated therapeutic processes, and the importance of considering the people's own perspective must guide one's actions when treating these people (Fig. 1.1).

Fig. 1.1 This drawing was by Tawaiku Juruna, an indigenous health agent. Body Paint Art from Yudja people, Xingu Indigenous Park, Brazil.

Suggested Reading

1. Buchillet D. A Antropologia da Doença e os Sistemas Oficiais de Saúde. In: Medicinas Tradicionais e Medicina Ocidental na Amazônia. Belém: Ed. Cejup; 1991.
2. Junqueira C. O Mundo Invisível. In: Revista Margem, vol. 10. São Paulo: Faculdade de Ciências Sociais, PUC-SP, Fapesp; 2000. p. 129–38.
3. Clastres P. Da Tortura nas Sociedades Primitivas. In:A Sociedade contra o Estado – Pesquisas de Antropologia Política. São Paulo: Cosac Naify; 2000.
4. Dalio J. Da Cultura do Corpo. Campinas: Papirus; 1995.
5. Instituto Socioambiental (ISA). Povos Indígenas no Brasil: Ikpeng. 2010. Disponível em http://pib.socioambiental.org/pt/ikpeng, acesso em mar.
6. Kaiabi; Aramut; Aarasi; Aruta; Kwariup; Pirapy; Tafarejup; Tamakari; Wyra'wat &
7. Yefuka. "Tradução do Mito de Origem dos Alimentos", documento elaborado durante o Curso dos Agentes de Saúde, nov; 2007.
8. Langdon EJ. Representações de Doenças e Itinerário Terapêutico dos Siona da Amazônia Colombiana. In: Saúde e Povos Indígenas. Rio de Janeiro: Ed. Fiocruz; 1994.
9. A Doença como Experiência: A Construção da Doença e seu Desafio para a Prática Médica. In: Conferência 30 Anos Xingu. São Paulo: Escola Paulista de Medicina (palestra); 1995.
10. Le Breton D. A Sociologia do Corpo. Petrópolis: Editora Vozes; 2006.
11. Mendonça, S. B. M. Relatório do Curso de Formação dos Agentes Indígenas de Saúde do Médio, Baixo e Leste Xingu – Módulo sobre Meio Ambiente, Doenças de Pele e Parasitoses Intestinais, jun. PIX; 2007.
12. Revista Margem, vol. 10. São Paulo: Faculdade de Ciências Sociais, PUC-SP/Fapesp; 1999. p. 129–38.
13. Le Breton D. A Sociologia do Corpo. 2a ed. Rio de Janeiro: Vozes; 2007. p. 39–61.
14. Clastres P. A Sociedade contra o Estado: Pesquisas de Antropologia Política. Cosac Naify: São Paulo; 2000.

Indigenous Peoples in Brazil

2

We know that the occupation of Brazil by paleoindians occurred more than 12,000 years ago. Migratory theories are known from Northeast Asia, using the strip of land called Beringia, which would have arisen as a result of glaciations that dropped sea levels by 50 m, with the migrants initially colonizing Alaska and North America and then taking the rest of the American continent.

However, the discovery of archaeological sites in the southeastern states of Piauí, Bahia, and Minas Gerais showed the evidence of human occupation in Brazil up to 25,000 years ago [7]. Thus, human presence in South America may date from the same period as in North America, suggesting that other forms of migration may have occurred in addition to ground migration through the Bering Strait and Alaska, such as crossing the Pacific in coastal waters and interisland navigation, similar to what would have occurred 50,000 years ago in the occupation of Australia.

Although numbers and migratory routes may differ, there is a consensus that the American continent became increasingly more densely populated with the arrival of Europeans. The estimated population of South America at the time of European arrival at the continent in 1492 is characterized by great variability: between 1 and 8.5 million inhabitants in the lowlands of South America. Some authors estimate between 1 and 6.8 million people lived in the Amazon, Central Brazil, and on the northeast coast, which would equate the population density of Brazil to that of the Iberian Peninsula in 1500.

Ethnologist Kurt Nimuendaju recorded about 1400 indigenous people – with major language families such as Tupi-Guarani, Jê, Aruak, Karib, Xirianá, Tukano – in the territory that corresponded to Brazil, recording an immense dispersion and diversity of peoples inhabiting Brazilian territory upon the arrival of the Portuguese.

The advance of colonization, so-called just wars,[1] enslavement, and extermination, as well as the spread of diseases such as smallpox, influenza, measles, and tuberculosis, caused high mortality rates among indigenous peoples. In the exchange of pathogens with colonizers and, later, with Africans brought as slaves by Portuguese, natives were the biggest losers since, except for certain fungal infections like *Tokelau* or *Tinha Imbricata*, nonserious or noncontagious diseases, and *treponematose*, known as *pinta* or *bouba*, no previously known diseases were transmitted to colonizers by indigenous peoples in Brazil.

It is argued that the main reason infectious diseases have had such an impact on indigenous communities is not necessarily that indigenous people lack specific genes related to immune responses, but the fact that Amerindian populations are biologically very homogeneous from a genetic point of view and unaware of infectious disease vectors that are widespread in Europe and that came over with the colonizers.

Mortality was of such magnitude that in five centuries disease reduced the original indigenous populations to just over 100,000 people, and it got to the point where some authors, such as anthropologist Darcy Ribeiro, called attention to the risk of indigenous peoples' extinction in Brazil in the 1960s and 1970s.

From the 1980s, contrary to expectations, a process of demographic recovery among indigenous peoples began, largely facilitated by newly acquired rights, especially as a result of the 1988 constitution, including the right to usufruct of traditional territories and access to health treatments and services.

Besides population growth, which can reach 4.5% per year in some groups owing to reductions in infant mortality and maintenance of high fertility rates, there is an emergency among groups considered extinct, especially in the Northeast, where they survived camouflaged among *cablocos* and *cafuzos*[2] because of persecution and prejudice.

[1] The Just Wars (Guerras Justas) were defined by 1570 Royal Charter, written by D. Sebastian, as legitimate procedures for the enslavement of indigenous peoples. According to Jacob Gorender, "just wars" were those authorized by the Crown and the colonial governors or those waged in self-defense against anthropophagous tribes' attacks. In Gorender [5].

[2] *Cafuzo* and *caboclo* refer to mixed races, the former between Africans and indigenous people, the latter between Europeans and indigenous people.

M.C. Florian et al., *Dermatological Atlas of Indigenous People*, DOI 10.1007/978-3-319-59446-0_2

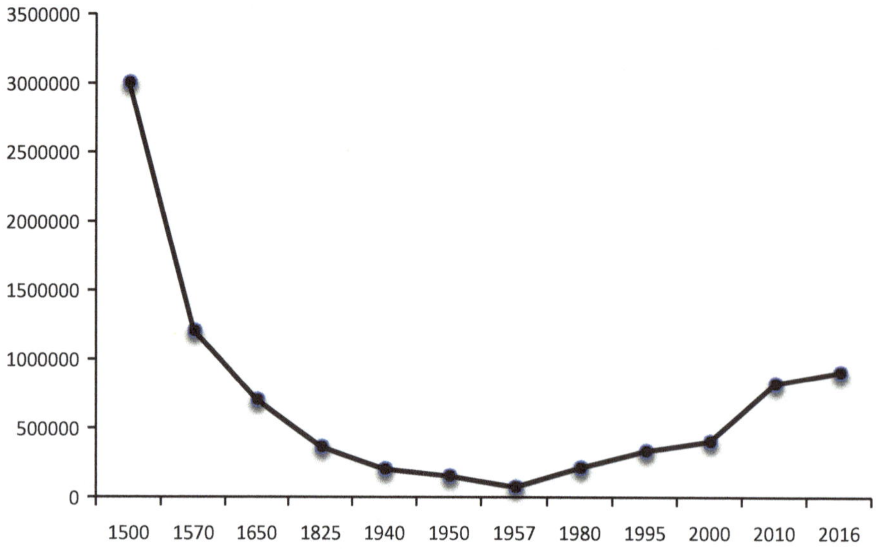

Death and demographic recovery of Indigenous peoples in Brazil, 1500–2016 (Source: Azevedo 2013; Brazilian Institute of Geography and Statistics, IBGE, 2010; Secretary of Special Indigenous Health, Ministry of Health – http://portalsaude.saude.gov.br/index.php/o-ministerio/principal/secretarias/secretaria-sesai/mais-sobre-sesai/9518-destaques. Accessed August 24, 2016)

Out of 817,963 people who identified themselves as indigenous in the last census, conducted in 2010, more than half lived in regularized indigenous lands (61.5%). Others lived in urban areas, in cities surrounding their traditional territories or state capitals, mostly in social exclusion in suburbs. The census also found 79,000 people who, though they considered themselves natives, chose not to identify themselves as such for various reasons, including prejudice, which remains intense in Brazil [9].

The distribution of indigenous lands in Brazil is very irregular and is related to the process that plays out following the discovery of wealth in different regions of the country. Thus, in the Northeast and Southeast, the first ones to be exploited, only 2% of lands are demarcated, and about 40% of the indigenous population lives in these regions. In the North and Midwest, especially in the Amazon, where exploitation began more recently, 98% of the land is demarcated, and 60% of the indigenous population lives in these areas.

There is wide dispersion in the 40% of indigenous population living in cities, as shown in the following table.

2.1 Sociocultural Diversity

Out of 1400 people registered as *Nimuendaju* in the early part of the last century, only 305 ethnic groups remain, and these groups speak 274 languages from various linguistic families. Also, according to the 2010 census, about 17.5% of the indigenous population does not speak Portuguese.

Despite all the tragedy that permeates relations between colonizers and indigenous societies, Brazil currently has the highest sociodiversity on the planet (Table 2.1).

Table 2.1 Distribution of indigenous populations living in urban areas, according to the Region and Federation Unit, Brazil, 2010

Region and federative unit	Population
North	**177,464**
Rondônia	6,707
Acre	8,976
Amazonas	95,215
Roraima	28,763
Pará	26,789
Amapá	3,770
Tocantins	7,244
Northeast	**115,215**
Maranhão	19,594
Piauí	1,366
Ceará	10,239
Rio Grande do Norte	1,272
Paraíba	12,489
Pernambuco	29,866
Alagoas	8,146
Sergipe	2,498
Bahia	29,745
Southeast	**47,704**
Minas Gerais	15,444
Espírito Santo	4,739
Rio de Janeiro	7,319
São Paulo	20,202
South	**39,499**
Paraná	13,251
Santa Catarina	9,241
Rio Grande do Sul	17,007
Miwest	**72,288**
Mato Grosso do Sul	38,971
Mato Grosso	26,513
Goiás	4,065
Distrito Federal	2,739

Source: Census 2010. IBGE

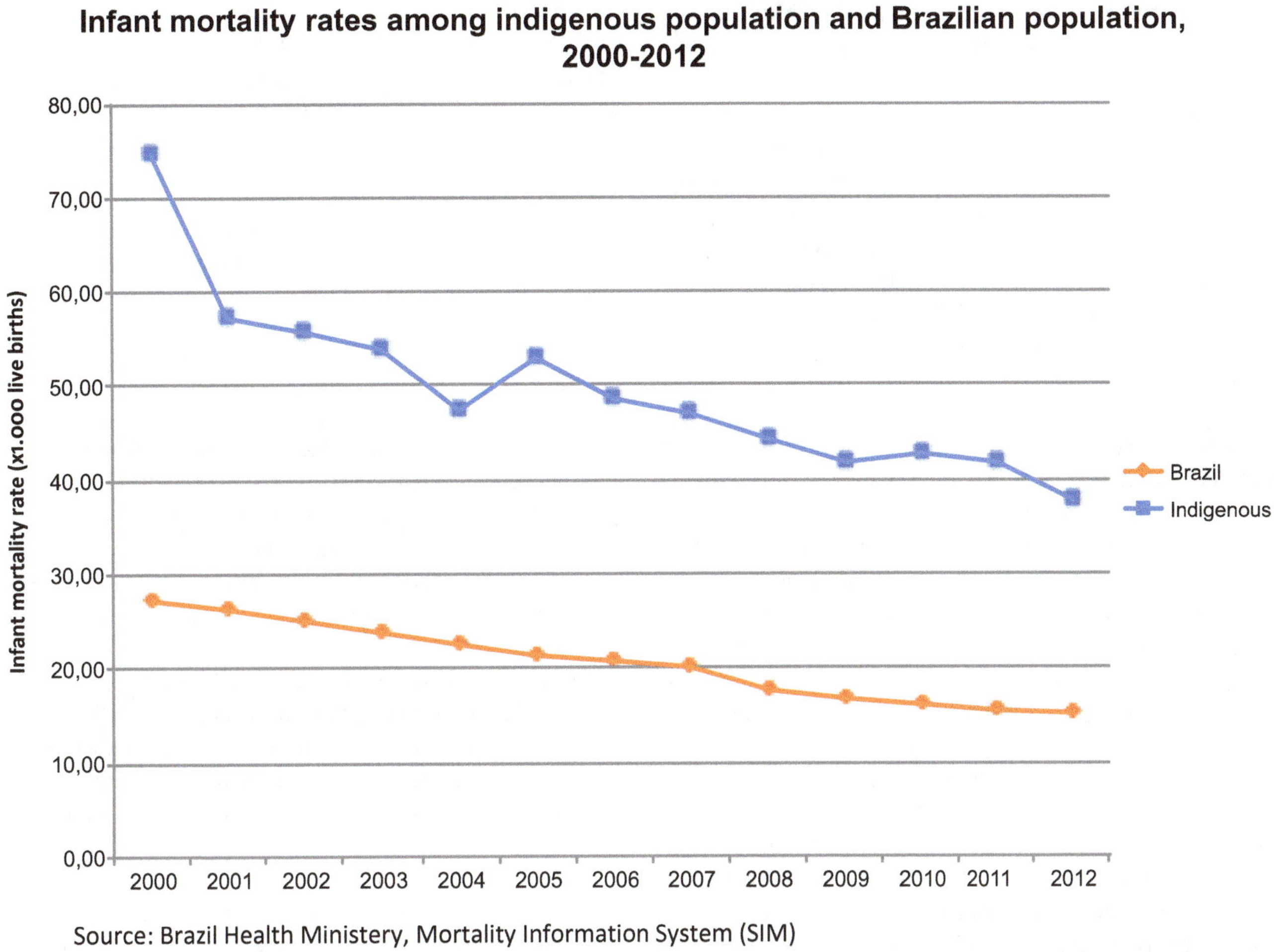

Infant mortality rates among indigenous population and Brazilian population, 2000–2012

2.2 Epidemiological Profile of Indigenous Peoples in Brazil

The socioeconomic status and epidemiological profile of indigenous peoples reflect their relationship with the larger national society. In general, infectious and parasitic diseases prevail, such as acute respiratory infections and gastroenteritis. In certain Amazon regions, endemic diseases like malaria and leishmaniasis have a high incidence among indigenous peoples, as is the case of the Yanomami in Roraima. In groups with more intense contact, rapid changes in lifestyle, especially the replacement of traditional food by processed products that are high in fat, sugar, and salt, and sedentariness, have been associated with chronic diseases, in particular metabolic syndrome, hypertension, and diabetes mellitus.

The epidemiological profile of indigenous peoples in Brazil is characterized by an epidemiological transition process. In this process, high rates of infectious and parasitic diseases coexist with a progressive increase in chronic diseases, including mental illness and suicide, and with abusive use of alcohol and illicit drugs in several communities. In general, the indicators of indigenous peoples' mortality are two to three times higher than among the Brazilian population as a whole. The following diagram illustrates the historical trend of infant mortality among indigenous peoples and Brazilians.

2.3 Uncontacted Indigenous Peoples or Groups in Brazil

Besides recognized indigenous peoples that also have regular contact with the national society, according to the *Fundação Nacional do Índio*, or National Indian Foundation (FUNAI), the agency responsible for indigenous policy, in 2013 there were 104 references to uncontacted indigenous groups in Brazil. Of these, 27 groups were confirmed by flyovers and incursions of FUNAI teams into their territory. The map below presents the locations of these references (*black* = not yet confirmed, *red* = confirmed) (Table 2.2).

Table 2.2 Information, from confirmed and unconfirmed sources, on uncontacted indigenous peoples in Brazil, 2013

Uncontacted indigenous records	Inside indigenous lands	Outside indigenous lands
Information	54	16
Unconfirmed sources	24	10
Confirmed sources	26	1

Source: CGIIRC/FUNAI, presentation at National Workshop on Health Protection and Promotion of Rights to Uncontacted and Recently Contacted Indigenous Peoples. Brasília, November 2013

A key measure of the protection of uncontacted Indians is to ensure their right to isolation, which is threatened by infrastructure projects in the Brazilian Amazon, especially the Initiative for Infrastructure Integration in South America (IIRSA).[3] Also in the South American macro context, another threat is represented by the Growth Acceleration Program (PAC), which aims to develop and integrate transport, energy, and communication infrastructure in Brazil. These projects aim to establish corridors (air, roads, rails, and waterways) to increase trade and establish supply chains directly related to the exploitation of natural renewable and nonrenewable resources. The supply chains are connected to global markets, mainly in North America, Europe, and Asia. In addition to these governmental projects, another threat to uncontacted groups are illegal activities, such as logging and gold and drug trafficking, that happen in border and gap areas in national South American territories.

It is important to emphasize that indigenous groups living in isolation achieve, over generations, a stable relationship with infectious disease agents present in their natural habitat, in a state of reasonable balance, allowing population growth. Contact with the surrounding society invariably introduces new disease agents, especially viral, which in the recent past have caused massive mortality (65, 66). Therefore, the advance of a market economy represents the biggest threat to the integrity of indigenous communities, which are seen as impediments to national economic growth.

Mistaken notions about of so-called primitive and obsolete societies contribute to this image and also reinforce integrationist proposals rejected by the Brazilian constitution, which guarantees indigenous peoples the right to live according to their customs and traditions, once cultural difference is permanent.

More recently, government income distribution, health, and education policies have reached the majority of indigenous communities in the country. However, although in some sectors, such as health care and education, indigenous peoples have excellent legal resources at their disposal, the incorporation of the right to be different is slow and difficult, as differentiated public policies must meet the real needs of indigenous peoples taking into consideration their enormous diversity.

The following figures show some aspects of indigenous life among different peoples (Figs. 2.1, 2.2, 2.3, 2.4, 2.5, 2.6, 2.7, 2.8, 2.9, 2.10, 2.11, 2.12, 2.13, 2.14, 2.15, 2.16, 2.17, 2.18, 2.19, 2.20, 2.21, 2.22, 2.23, 2.24, 2.25, 2.26, 2.27, 2.28, 2.29, 2.30, 2.31, 2.32, 2.33, 2.34, 2.35, and 2.36).

[3]Launched in September 2000, during the meeting of the 12 South American presidents and 350 Latin American business people held in Brasilia, the Inter-American Development Bank (BID) presented an action plan on the integration of South American infrastructure; available at http://www.oei.es/oeivirt/cimeira2.htm.

Fig. 2.1 Houses

Fig. 2.2 Village environment

Fig. 2.3 Indigenous people in a canoe on the Xingu River

Fig. 2.4 Aspects of indigenous life

Fig. 2.5 Indigenous children on a river

Fig. 2.6 Preparation of *beiju*, a flour extracted from cassava root

Fig. 2.7 Preparation of *beiju*, a flour extracted from cassava root

Fig. 2.8 Preparation of *beiju*, a flour extracted from cassava root

Fig. 2.9 Indigenous food

Fig. 2.11 Aspects of indigenous life

Fig. 2.10 Indigenous food

Fig. 2.12 Indigenous people

Fig. 2.13 Indigenous children on river

Fig. 2.14 Indigenous children on river

Fig. 2.15 Indigenous children on river

Fig. 2.16 Indigenous games

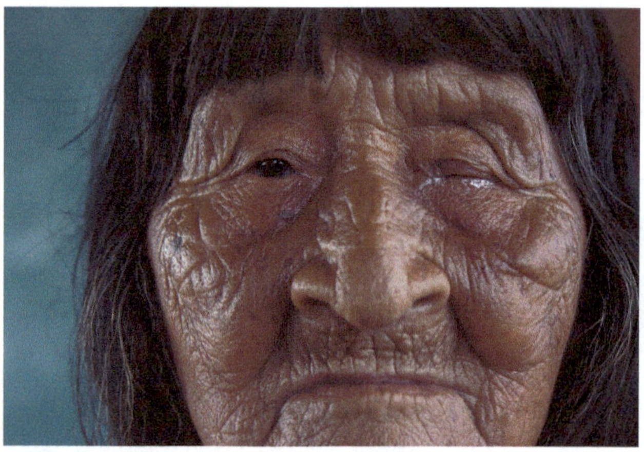

Fig. 2.17 Indigenous people

Fig. 2.18 Indigenous people

Fig. 2.19 Indigenous people

Fig. 2.20 Village environment

Fig. 2.21 Indigenous crafts

Fig. 2.22 Aspects of indigenous life

Fig. 2.23 Village environment

Fig. 2.24 Aspects of indigenous life

Fig. 2.25 Village environment

Fig. 2.26 Aspects of indigenous life

Fig. 2.27 Houses

Fig. 2.28 Indigenous people

Fig. 2.29 Indigenous people in a canoe on the Xingu River

Fig. 2.30 Village environment

Fig. 2.31 Village environment

Fig. 2.32 Village environment

Fig. 2.33 Indigenous people

Fig. 2.34 Village environment

Fig. 2.35 Aspects of indigenous life

Fig. 2.36 Village environment

Suggested Reading

1. Meltzer DJ. Why don't we know when the first people came to North America? Am Antiq. 1989;54(3):471–90.
2. Sapper, Karl (1924); Kroeber, Alfred (1934); Rosemblat (1954); Borah (1964); Dobyns (1966); Chaunu (1969); Denevan (1976); Hemming, John (1978). In: Handbookof American Indians, Smithsonian Institute.
3. Carneiro MC. Política Indigenista no Século XIX. In: Cunha MC org. História dos Índios do Brasil. São Paulo: Cia das Letras; 1992. p. 133–54.
4. Nimuendaju C. Mapa Etno Histórico de /IBGE edição fac-similar. Rio de Janeiro: IBGE [Brasília-DF]: Ministério da Educação. 2002. p. 94.
5. Gorender J. "O escravismo Colonial" (The colonial slavery). São Paulo: Ática; 1985. p. 490–1.
6. Rodrigues D. Saúde e doença entre os Panará, Povo Indígena Amazônico de contato recente, 1975–2007. [Tese]. UNIFESP. 2013.
7. Pagliaro H, Azevedo MM , Santos RV. (Org). Demografia dos povoindígenas no Brasil. Rio de Janeiro: Ed. FIOCRUZ; Campinas: Associação Brasileira de Estudos Populacionais – ABEP. 2005. p. 192. (Coleção Saúde dos povos indígenas).
8. Pereira NOM, Azevedo MM, Santos RV. Perfil demográfico esocio-econômico das pessoas que se autodeclararam indígenas nos censos demográficosde 1991 e 2000. In: PAGLIARO H, AZEVEDO MM, SANTOS RV. (Org.). Demografia dos povos indígenas no Brasil. Rio de Janeiro: Ed. FIOCRUZ; Campinas:Associação Brasileira de Estudos Populacionais – ABEP. 2005. p. 155–66. (ColeçãoSaúde dos povos indígenas).
9. IBGE. O Brasil Indígena no Censo populacional de 2010. Available in http://indigenas.ibge.gov.br/. Accessed in 09/01/2016.
10. Vaz A. Isolados do Brasil, Informa IWGIA n.10. Available in http://www.iwgia.org/iwgia_files_publications_files/0453_informe_10_ne.pdf. Accessed in 08/24/2016.

Elementary Skin Lesions in Dermatological Medical Examinations

3

The recognition of elementary skin lesions is the most important part of clinical examinations for skin disease diagnosis and for indicating the best treatment. We present a sample of some important elementary lesions in skin examinations found in most indigenous peoples, according to our observations in fieldwork.

3.1 Papule

A small (less than 1.0 cm) raised lesion with solid content (Fig. 3.1).

3.2 Plaque

A raised lesion with solid content, larger than 1.0 cm, generally with flat surface (Fig. 3.2).

3.3 Nodule

Lesion with solid content and round shape that may present as a rise on the skin or as a deeper injury, with size larger than 1.0 cm (Fig. 3.3).

3.4 Verrucous or Wartlike Lesions

A lesion with solid content and yellowish-white irregular surface (Fig. 3.4).

3.5 Vesicle and Bullae

A lesion with liquid and clear content measuring less than 1.0 cm is called a vesicle. A lesion with liquid content measuring more than 1.0 cm is called a bulla (Fig. 3.5).

3.6 Pustule

Lesion with purulent content (yellowish) measuring less than 1.0 cm (Fig. 3.6).

3.7 Erythematous Macule

Change in color, usually red, which disappears under finger pressure (Fig. 3.7).

3.8 Hypochromic Macule and Achromic Macule

Change in skin color that is lighter than normal skin (hypochromia) (Fig. 3.8) or completely white (achromia) (Fig. 3.9).

3.9 Hyperchromic Macule

Change in skin color that is darker than normal skin (Fig. 3.10).

3.10 Edema

Leakage of fluid in skin layers giving bloated look. It can present with the natural color of skin or like erythema (Fig. 3.11).

3.11 Ulcer

Loss of skin continuity solution – dermis and epidermis (Fig. 3.12). When the lesion is superficial (epidermis) it is called erosion.

M.C. Florian et al., *Dermatological Atlas of Indigenous People*, DOI 10.1007/978-3-319-59446-0_3

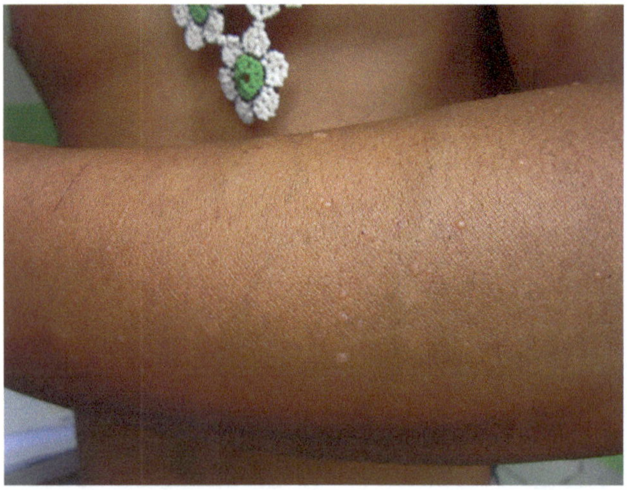

Fig. 3.1 Papules on forearm

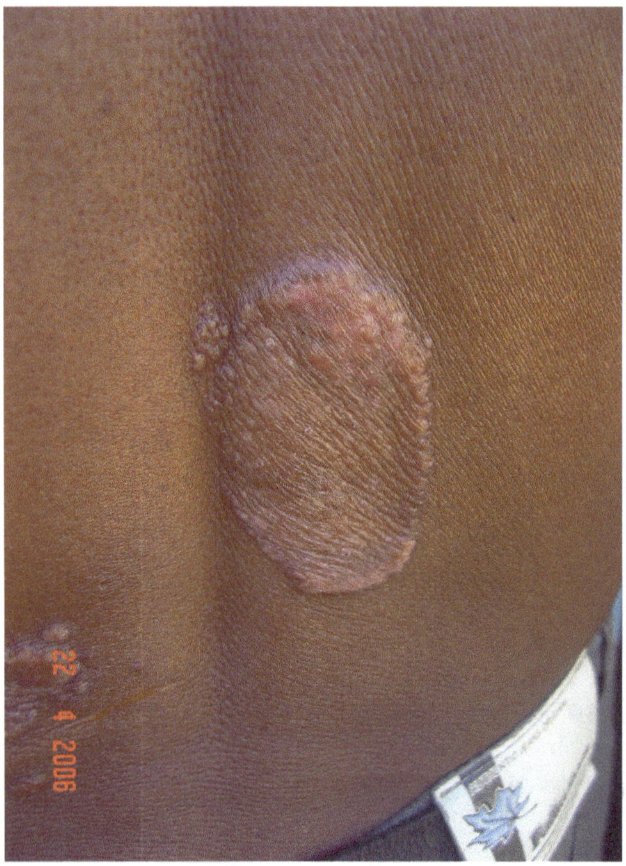

Fig. 3.2 Plaque on dorsum

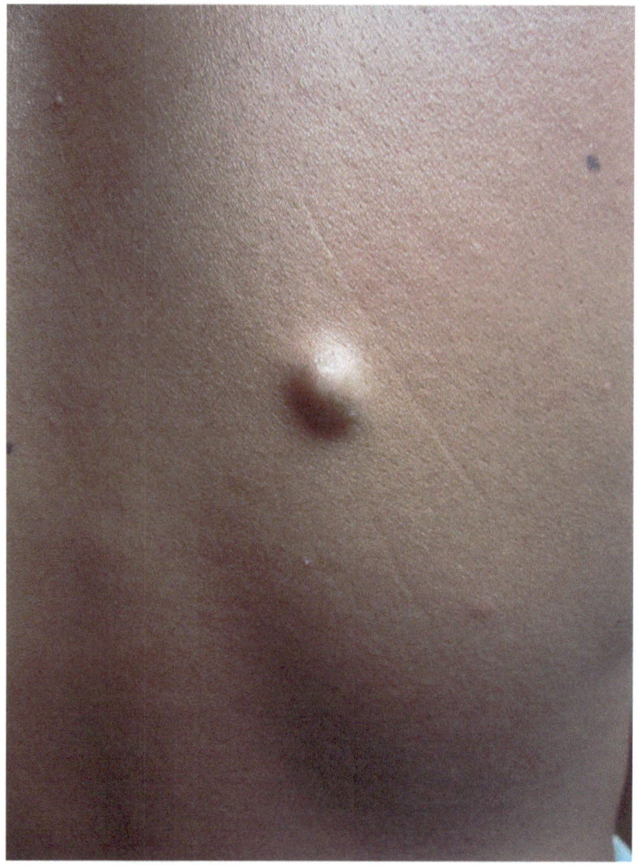

Fig. 3.3 Nodule on dorsum

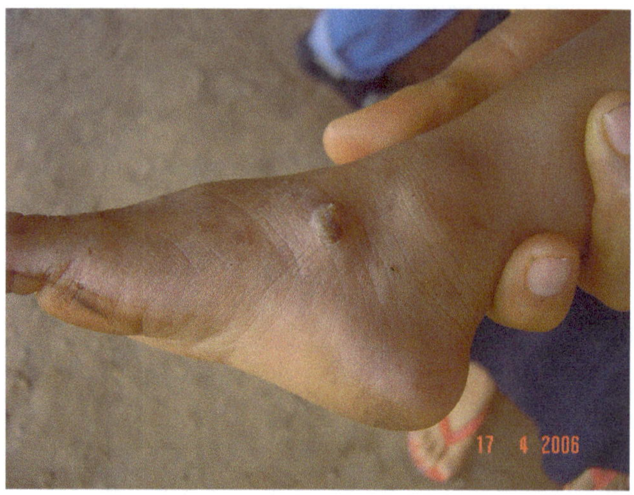

Fig. 3.4 Wartlike lesion on foot

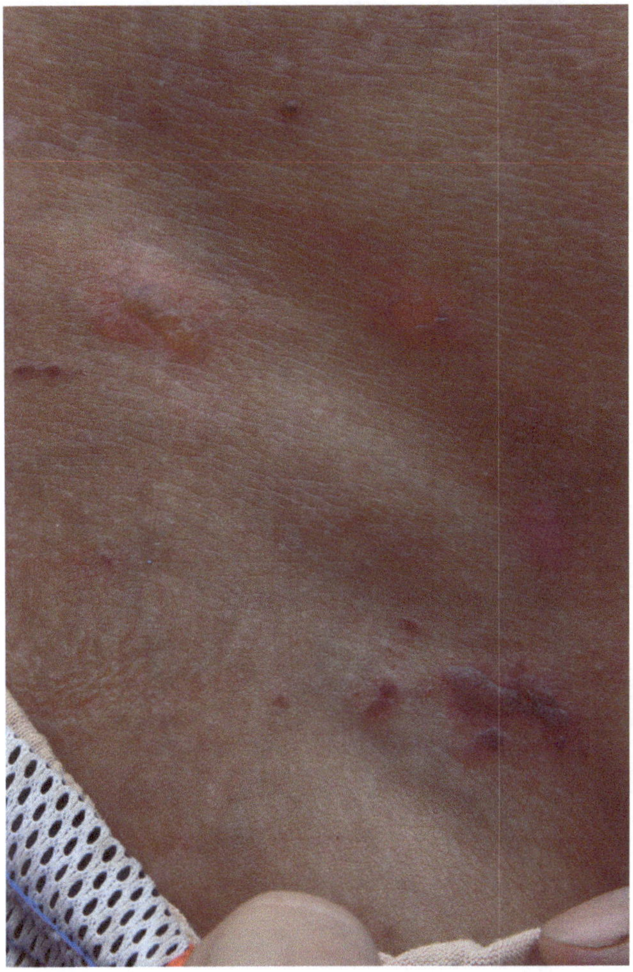

Fig. 3.5 Vesicles and bullae

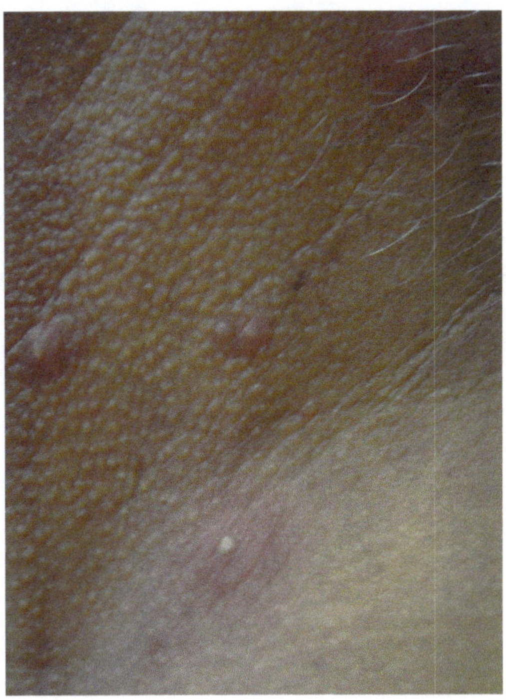

Fig. 3.6 Pustule

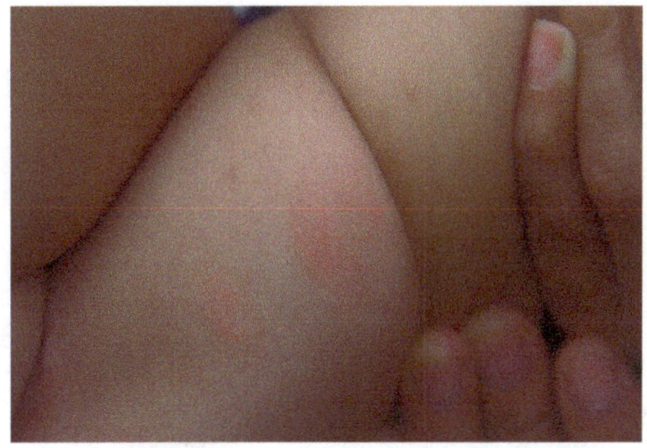

Fig. 3.7 Erythematous macules on thigh

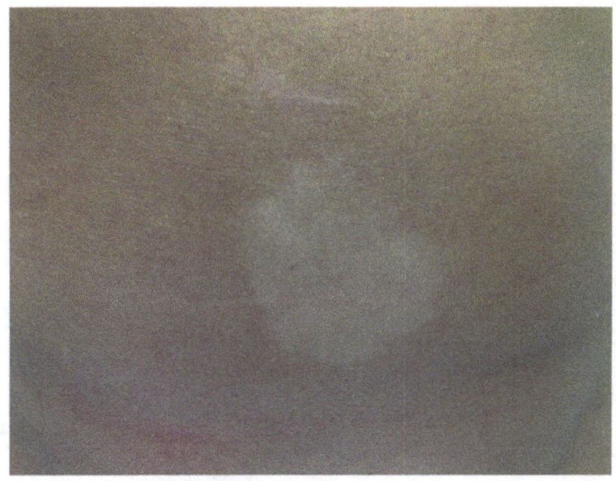

Fig. 3.8 Hypochromic macule on abdomen

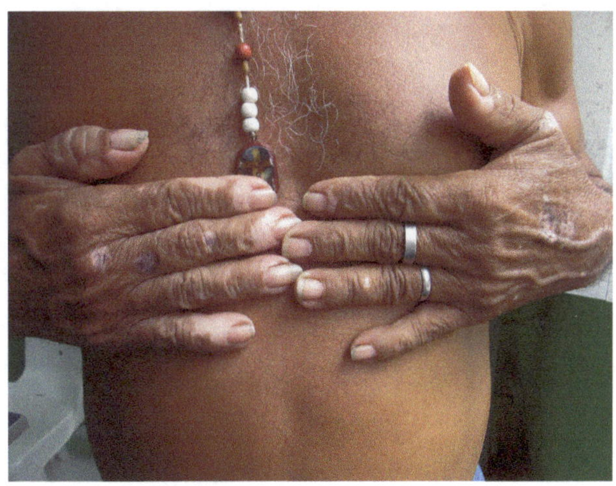

Fig. 3.9 Achromic macules on hands

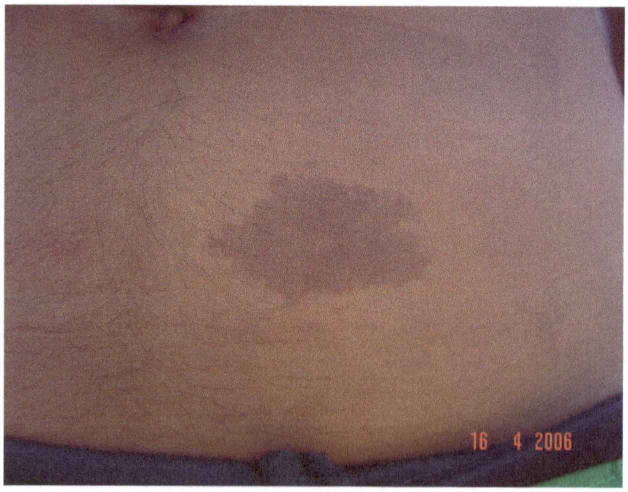

Fig. 3.10 Hyperchromic macule on abdomen

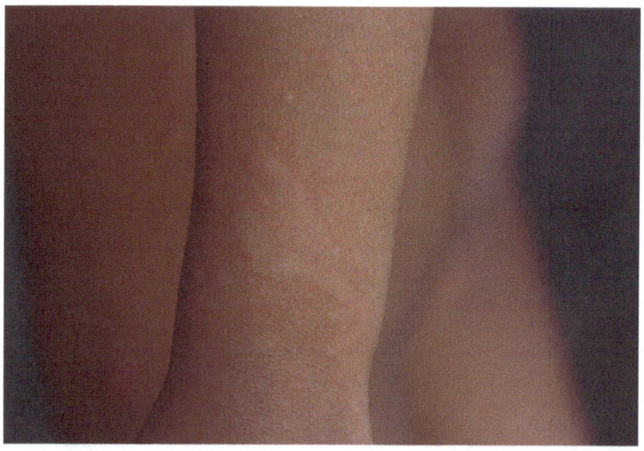

Fig. 3.11 Edema on forearm

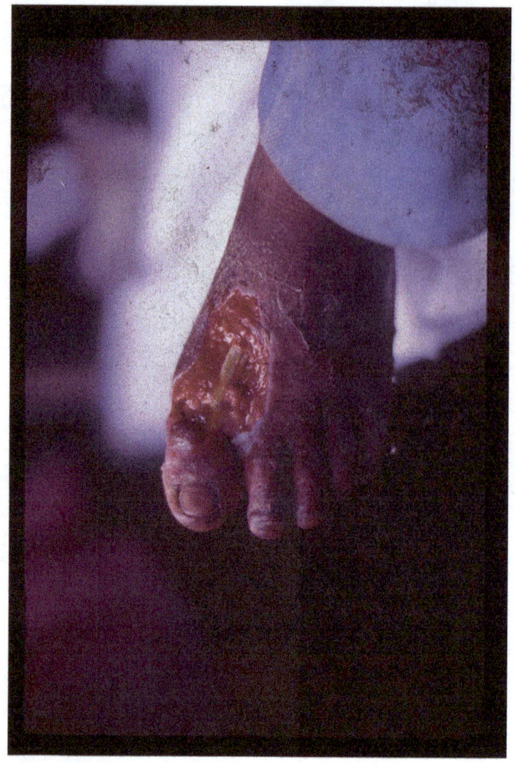

Fig. 3.12 Ulcer on foot

Fig. 3.13 Scales on leg

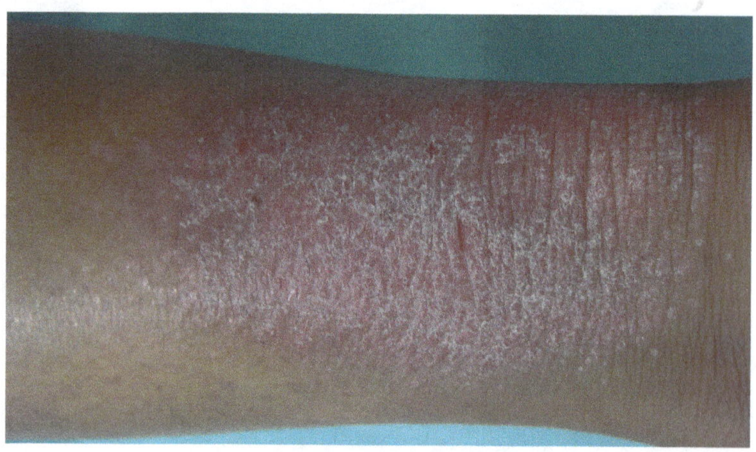

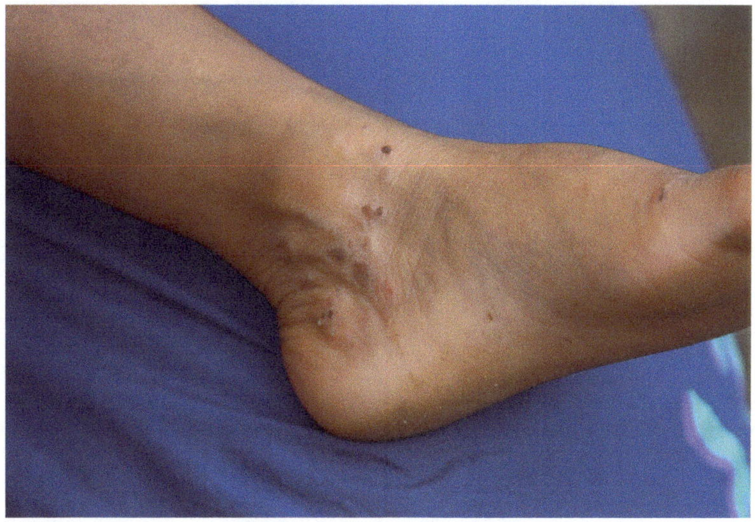

Fig. 3.14 Crusts on ankle

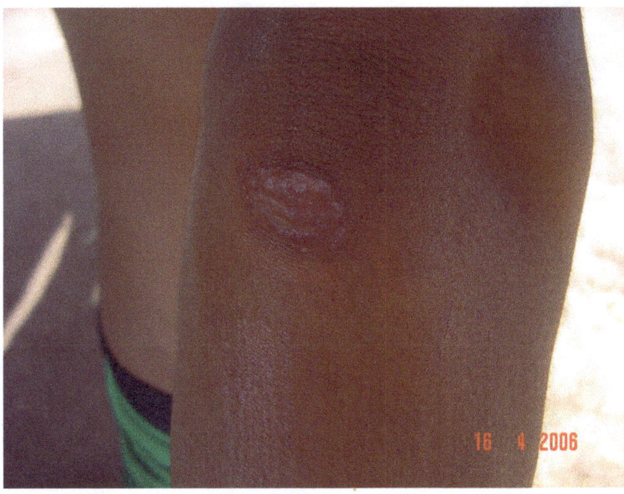

Fig. 3.15 Atrophy on elbow

3.12 Scales

Loss of superficial skin layers (peeling) (Fig. 3.13).

3.13 Crusts

Lesions that cover areas in which there is a loss of skin continuity solution (ulcers or broken blisters). They serve as a kind of biological dressing (Fig. 3.14). The crust can be the color of honey, yellowish or reddish brown (hematic).

3.14 Atrophy

Areas with skin thin aspect because loss of tissue (epidermis, dermis, or subcutaneous) (see Fig. 3.15).

Suggested Reading

1. Griffiths C, Barker J, Bleiker T, Chalmers R, Creamer D. Rook's textbook of dermatology. 9th ed. Chichester/West Sussex: Wiley; 2016.
2. Rotta O. Dermatologia. São Paulo: Ed Manole; 2008.
3. Goldsmith LA, Katz SI, Gilchrest BA, Paller AS, Leffell DJ, Wolff K. Fitzpatrick's dermatology in general medicine. 8th ed. New York: McGraw-Hill; 2012.

Infectious Diseases

4

4.1 Fungal Diseases

4.1.1 Dermatophytosis

Fungal diseases are diseases that affect the skin and appendages (hair and nails). They are caused by fungi known as dermatophytes, which may be anthropophilic, zoophilic, or geophilic. One of the characteristics of these fungi is their affinity for keratin present on the skin, hair, and nails.

These frequent diseases are common in children and adults. Transmission can occur through personal contact, contact with animals (dogs, cats), and contact with the ground.

The most frequent causative agent is *Trichophyton rubrum*, which causes lesions on the nails, feet, and groin. Older people are most affected. *Microsporum canis* is another frequent etiological agent. It infects animals like cats (which can be only carriers) or dogs (which in general experience hair loss).

The infection can affects any site in the body, including hair and nails. Dermatophytoses go by different names based on the area of the body affected: tinea corporis, tinea pedis, tinea capitis, tinea inguinalis, and onycomychosis.

Tinea corporis presents with itchy lesions as well as erythematous and scaly well-defined borders. These lesions might have an annular shape and may be single or multiple (Figs. 4.1 and 4.2). Tinea pedis is common in individuals who use closed shoes and has been related to increased foot sweating. Lesions may affect the interdigital region (Fig. 4.3). Bromidrosis, itching, and local maceration are part of the clinical presentation. Dry desquamation of the foot occurs in the chronic form. Tinea capitis presents as hair loss (alopecia) and squamous skin in these areas. It is common in children. Tinea inguinalis is more common in men, occurs in the warmer months, and has been related to the use of clothes and local sweating. The lesions are erythematous and scaly, and there is intense itching. In onychomycosis, there is a detachment of the nail that evolves with thickening of the nail plate. Nail dyschromia (yellowish or blackened nails) is also common (Fig. 4.4). There is a form of mycosis in which the fungus is located deeply in the skin, causing folliculitis and perifolliculitis. It is called *granuloma trichophyticum*. Clinically the lesions are erythematous papules or papule pustules and mainly affect the extremities or face.

For diagnostic confirmation a direct mycological examination is performed. By scraping the edge of the lesions with a blade skin scales are obtained that are placed on a glass slide. After adding a drop of solution of KOH (potassium hydroxide) at 10%, a reading can be obtained in an optical microscope (Figs. 4.5 and 4.6). The fungus culture in the appropriate medium determines the species of fungus.

The treatment will be selected according to the extent of the disease, its location, the patient's clinical condition, and the causative agent. As a rule, treatment with topical medications is sufficient (clotrimazole, miconazole, terbinafine, amorolfine, ketoconazole, and others). The use of systemic antifungals (fluconazole, itraconazole, griseofulvin, terbinafine) is indicated in cases of extensive lesions in the skin.

4.1.2 Paronychia

Paronychia is a common disease affecting mainly adults whose hands are continually exposed to moisture (water). *Candida* spp. Fungi are involved in the pathogenesis. Association with *Staphylococcus* spp. is common. The clinical presentation is a swelling of the proximal nail region associated with erythema. When chronic, the area is hardened (Figs. 4.7 and 4.8) and the nail becomes irregular, yellow, and thickened. If there is secondary bacterial infection, the region becomes swollen again, in addition to experiencing local pain and purulent discharge. For diagnosis confirmation, a direct mycological examination involving skin scraping (scales) can be carried out. However, a secondary bacterial infection is accompanied by a lower direct mycological positivity.

M.C. Florian et al., *Dermatological Atlas of Indigenous People*, DOI 10.1007/978-3-319-59446-0_4

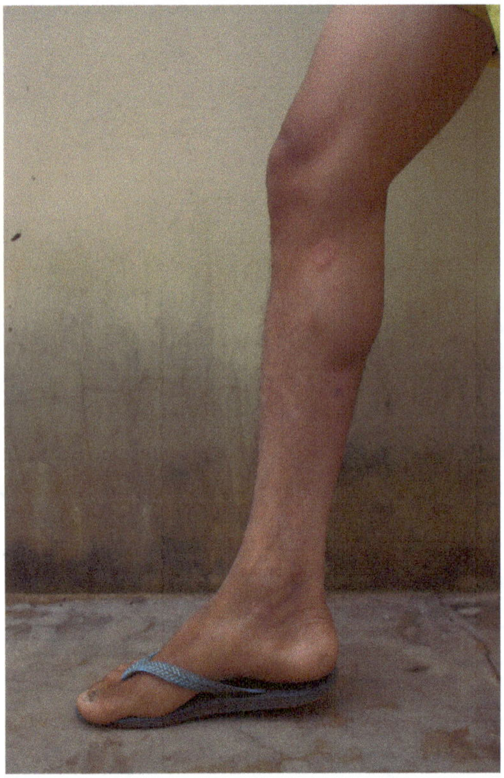

Fig. 4.1 Tinea corporis: lesion on leg

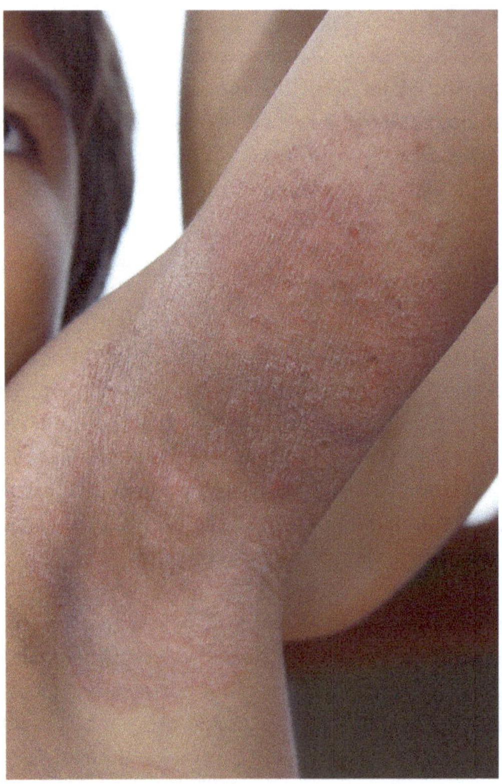

Fig. 4.2 Tinea corporis: lesion on armpit

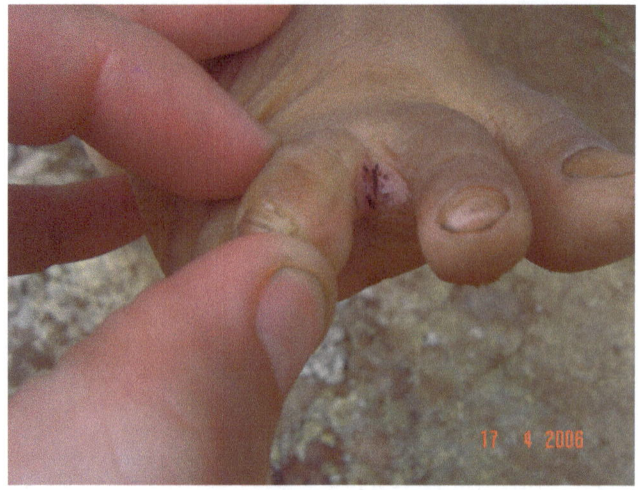

Fig. 4.3 Tinea pedis: lesions on interdigital region

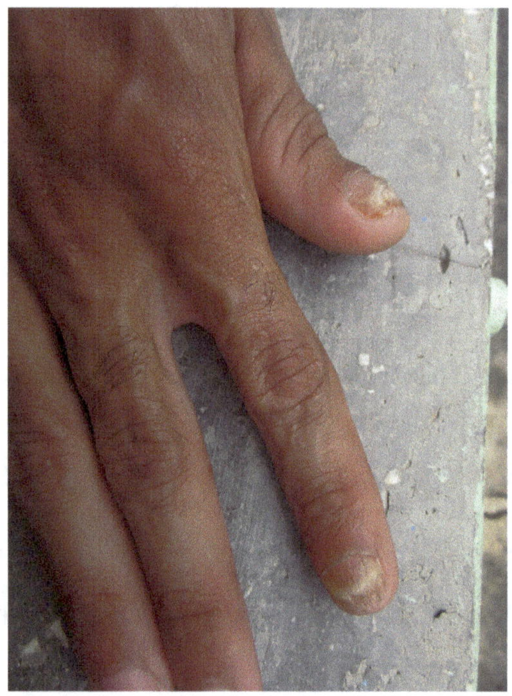

Fig. 4.4 Onichomycosis: nail dystrophy

Usually treatment with topical medications is sufficient. Nystatin or azoles (ketoconazole, miconazole, clotrimazole) should be applied twice daily until clinical cure occurs. In case of a secondary bacterial infection, topical antibiotics (gentamicin, neomycin, mupirocin, or fusidic acid cream) should be used. In some cases a combination of topical corticosteroids with topical antibiotics may be used (betamethasone/gentamicin, betamethasone/fusidic acid). In rare cases, systemic drugs are necessary.

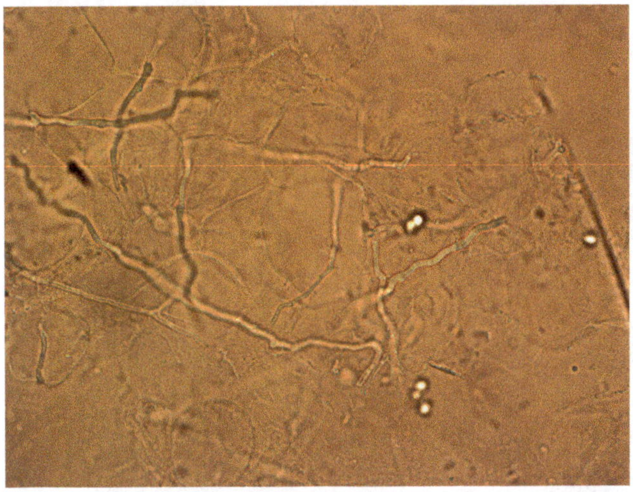

Fig. 4.5 Dermatophytosis: direct mycological examination of scales (40×) (Photo by Orion Sant Ana Motter Borba)

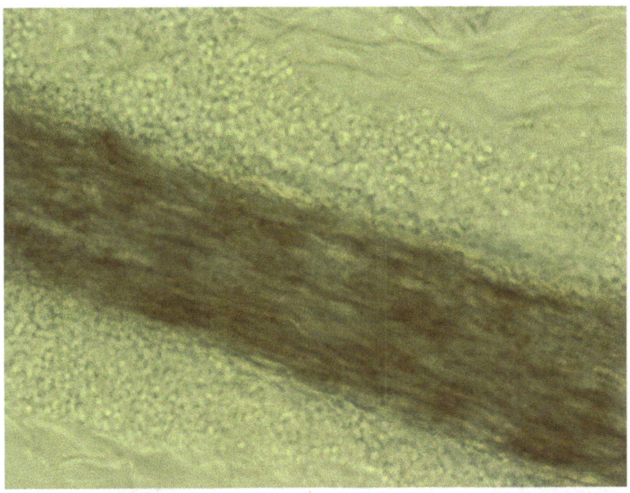

Fig. 4.6 Dermatophytosis: direct mycological examination of hair (40×) (Photo by Orion Sant Ana Motter Borba)

4.1.3 Pityriasis Versicolor

Pityriasis versicolor is a superficial fungal infection caused by *Malassezia* spp. Fungi. It is a common disease that affects adults and, rarely, children before puberty. However, it is not highly transmissible like other superficial mycoses. It occurs more often in tropical and subtropical regions during hot and humid months. Lesions become more apparent after sunlight exposure due to a color contrast because the parasitized site does not tan. The clinical presentation is white, erythematous, or brownish macules, usually round and with fine scaling. The macules usually occur on the back, chest, neck, face, and arms (Figs. 4.9, 4.10, and 4.11) and are asymptomatic.

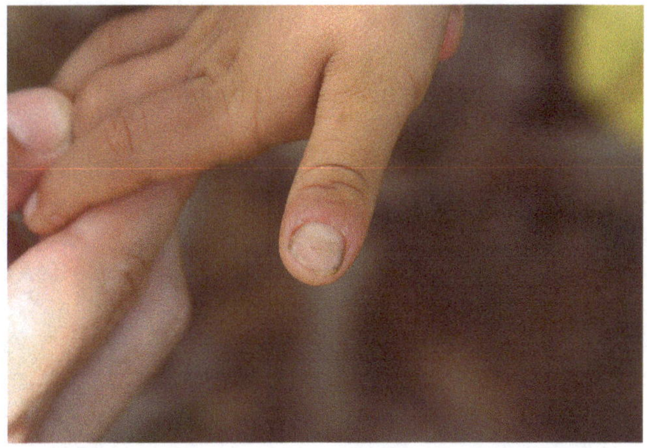

Fig. 4.7 Paronychia: lesions on proximal nail region

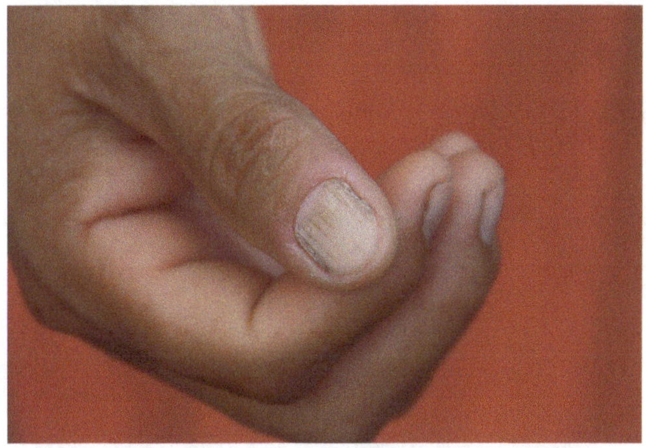

Fig. 4.8 Paronychia: lesions on proximal nail region

When necessary, a direct microscopic examination by skin scraping (scales) can confirm the diagnosis (Fig. 4.12). Fungal culture or molecular biology tests can identify the fungal species. Treatment with topical medications (clotrimazole, miconazole, terbinafine, amorolfine, ketoconazole, and others) is sufficient in most cases. Shampoos (ketoconazole, ciclopirox olamine, zinc pyrithione, selenium sulfide) are effective too. If the affected area of the skin is large, systemic drugs are indicated (fluconazole in weekly doses or itraconazole).

4.1.4 Tinea Imbricata

Tinea imbricata is a rare fungal infection caused by *Trichophyton concentricum*.

It occurs in tropical areas and has been associated with the Xingu Indigenous Park (Brazil). In some indigenous languages it is known as *tokelau* or *chimberê*. The main charac-

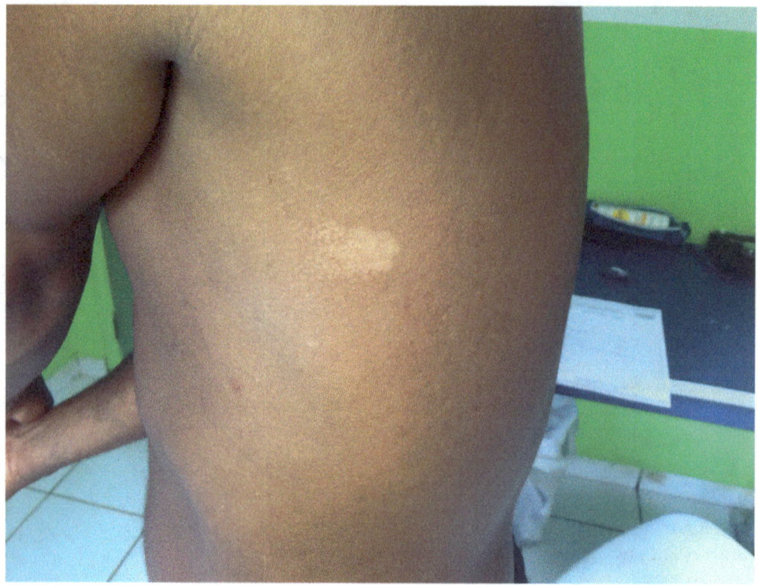

Fig. 4.9 Pityriasis versicolor: lesions on dorsum

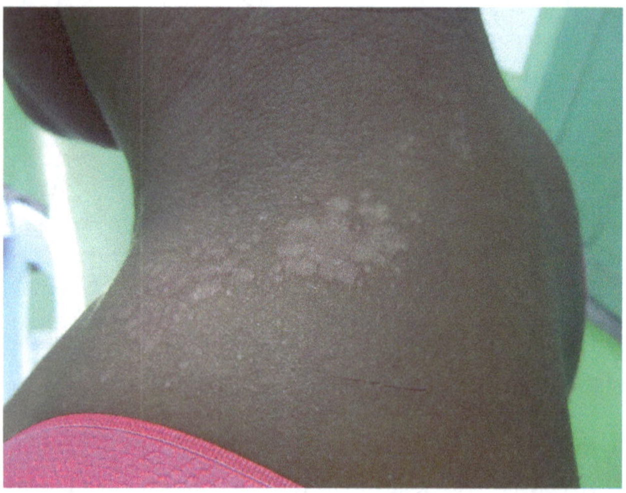

Fig. 4.10 Pityriasis versicolor: lesions on dorsum

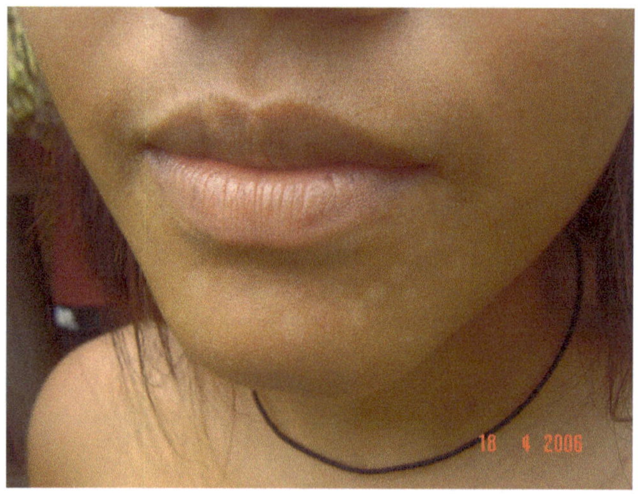

Fig. 4.11 Pityriasis versicolor: lesions on face

teristic is the presence of scaly lesions in a concentric arrangement (in regular circles), located on the trunk, arms, and legs, that can affect a large area of the skin (Figs. 4.13 and 4.14). Direct microscopic examination by skin scraping (scales) can be used to confirm the fungus's presence. However, the fungal species can be determined with certainty only in culture. Treatment is carried out with the same topical medications that are used to treat dermatophytoses (clotrimazole, miconazole, terbinafine, amorolfine, ketoconazole, and others). The use of systemic antifungals (fluconazole, itraconazole) is indicated when a large area of skin is affected.

4.1.5 Jorge Lobo's Disease

Jorge Lobo's disease affects the skin and subcutaneous tissue and is caused by the fungus *Lacazia loboi. The* Caiabi Indians call it *miraip*. Only a few hundred cases have been reported in the world. It is found mainly in the Amazon region in individuals who frequent the forests (e.g., hunters). There is a large number of cases among the Caiabi Indians. The path of transmission is not known. However, among the Kayapo, it is believed that the disease was acquired in its original territory 300 km to the north of the Xingu Indigenous Park (XIP). Many Caiabi Indians have migrated to the XIP in the last 40 years, and no new cases of Jorge Lobo's disease

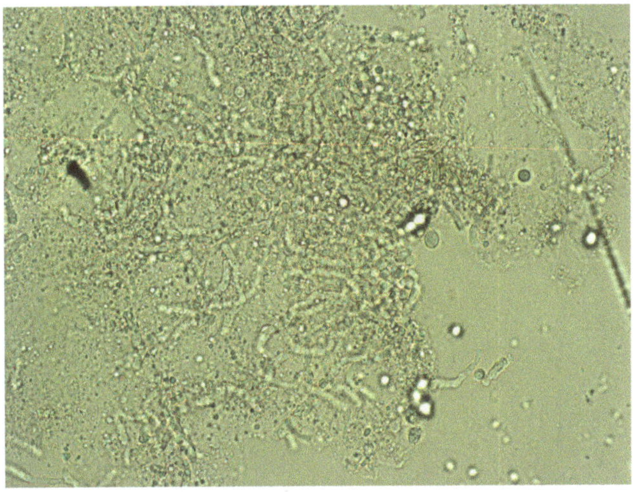

Fig. 4.12 Pityriasis versicolor: direct mycological examination of scales (40×) (Photo by Orion Sant Ana Motter Borba)

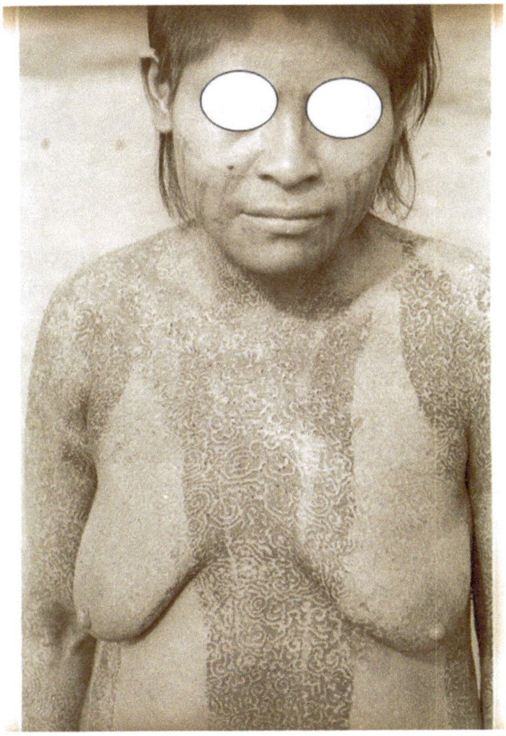

Fig. 4.13 Tinea imbricate: disseminated lesions

native to XIP has emerged. We know that the disease is probably not transmitted from person to person. In addition to humans, the disease also affects some species of cetaceans (*Tursiops truncatus* and *Sotalia guianensis*).

The clinical presentation of Jorge Lobo's disease is polymorphic. Many types of skin lesions are observed: reddish color and hardened nodules or plaques, atrophic lesions, and keloid-like lesions. The lesions can affect any body region and may be single or multiple (Figs. 4.15, 4.16, 4.17, 4.18,

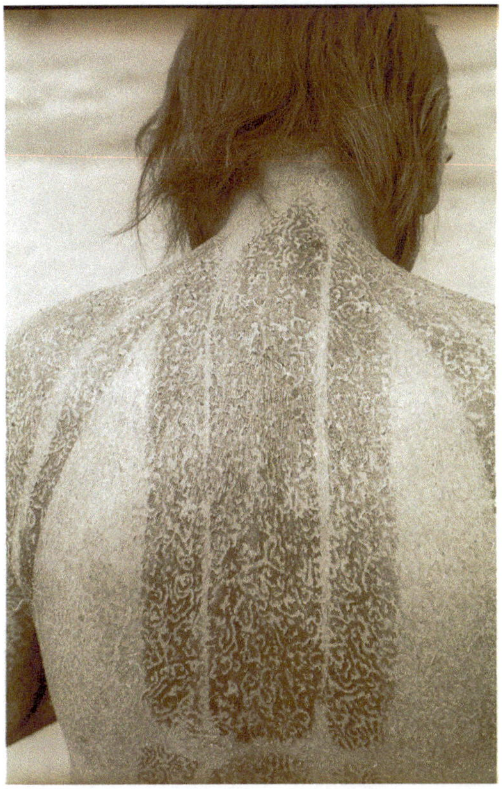

Fig. 4.14 Tinea imbricate: disseminated lesions

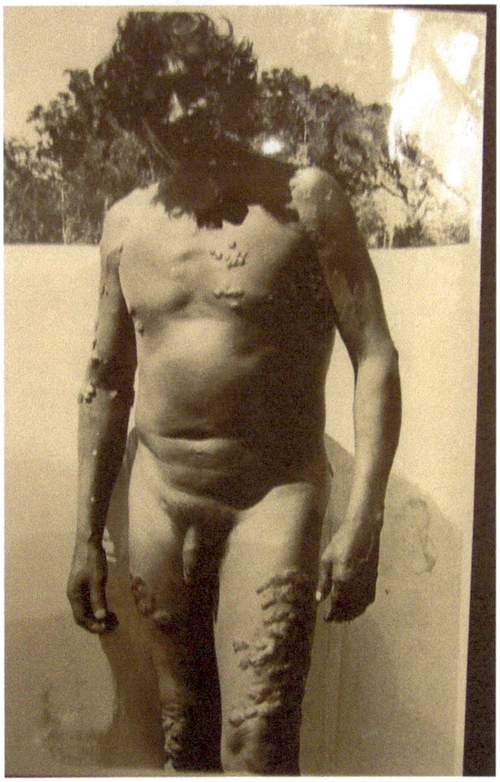

Fig. 4.15 Jorge Lobo's disease: multiple lesions

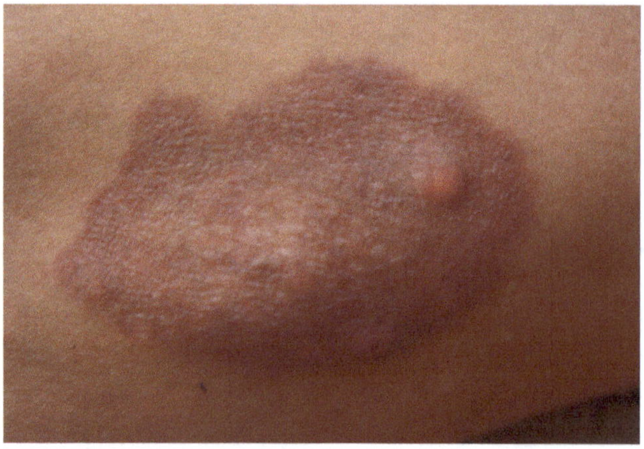

Fig. 4.16 Jorge Lobo's disease: single lesion

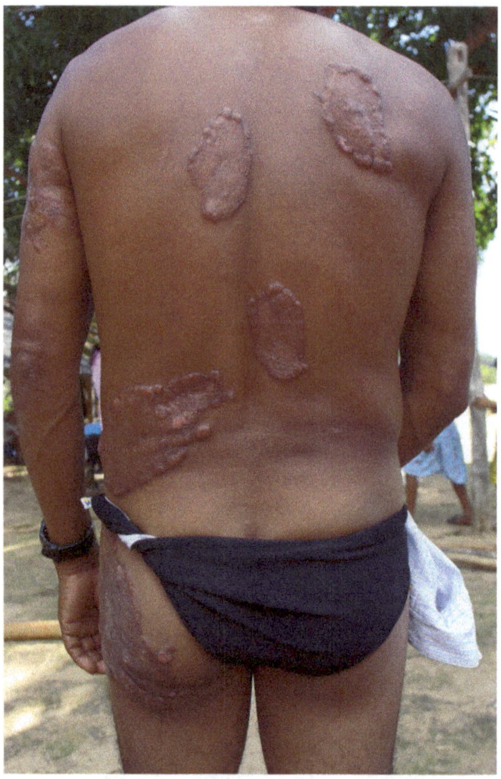

Fig. 4.17 Jorge Lobo's disease: multiple lesions

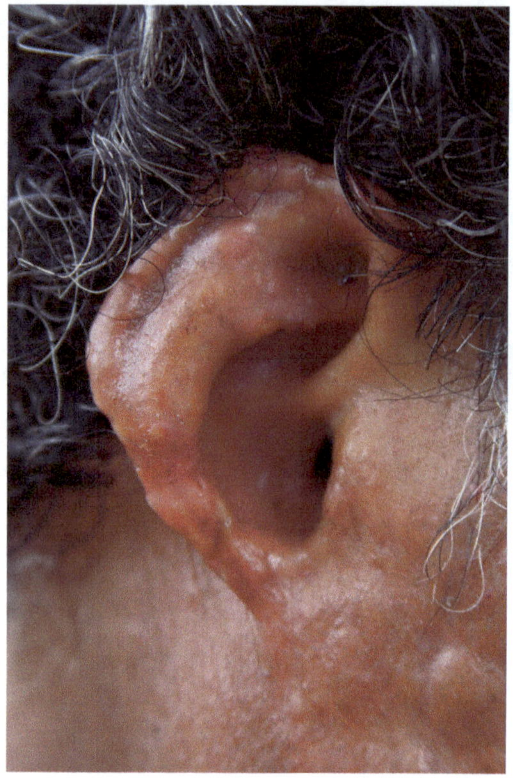

Fig. 4.18 Jorge Lobo's disease: destructive lesions

and 4.19). Ulcers can be observed in nodules as in atrophic lesions (Fig. 4.20). The overall health of the patient is not affected unless there are complications such as bacterial infections. The transformation into squamous-cell carcinoma is unusual, but it can occur.

Diagnosis is made by direct microscopic examination. A pathological examination will show the presence of lymphocytic infiltrate and round fungal structures with birefringent walls, isolated or grouped in the form of chains (Figs. 4.21 and 4.22). To date, no one can grow the *Lacazia loboi* fungus in culture medium. There is an antigenic resemblance between the fungi *Lacazia loboi* and *Paracoccidioides braziliensis*.

Treatment is difficult because there are no known effective drugs against the disease. There are reports of the use of systemic drugs like ketoconazole, itraconazole, posaconazole, clofazimine, amphotericin B, and 5-fluorocytosine, as well as a multidrug therapy for multibacillary leprosy with little in the way of results. In some cases the surgical removal of lesions may heal the patient, especially if it done early (at the initial stage and when there are few lesions).

4.1.6 Chromoblastomycosis

These mycoses, which are caused by a group of dark fungi, affect the skin and subcutaneous tissue . The most common etiological agents change depending on the geographic region. The most common agent in Brazil is *Fonsecaea pedrosoi*. These diseases occur in tropical countries, especially in rural areas. The infection occurs by a trauma to the skin where the fungus penetrates; thus, it is a mycosis of implantation. Local fungus development will occur depending on the immune status of the host.

The emergence of papules or nodules that develop into a large verrucous plaque is the primary clinical manifestation (Figs. 4.23 and 4.24). The presence of black dots on the

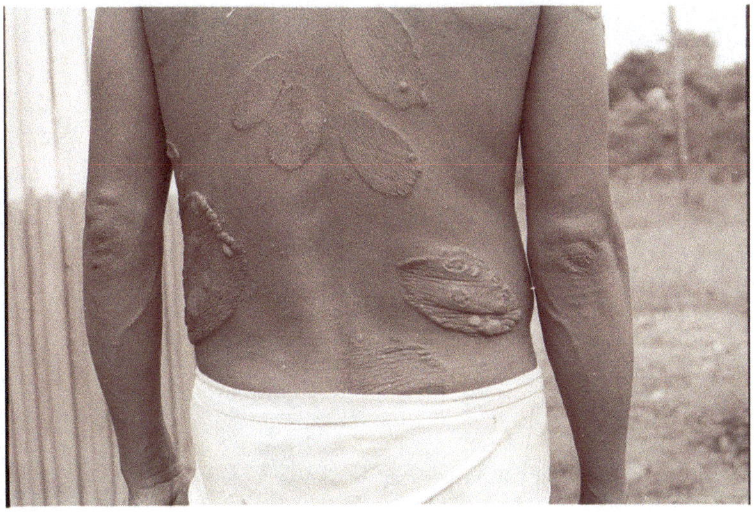

Fig. 4.19 Jorge Lobo's disease: multiple lesions

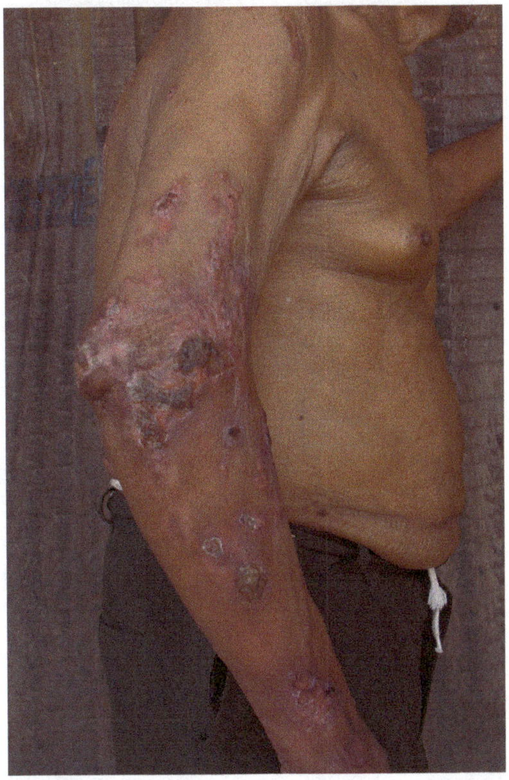

Fig. 4.20 Jorge Lobo's disease: ulcers lesions

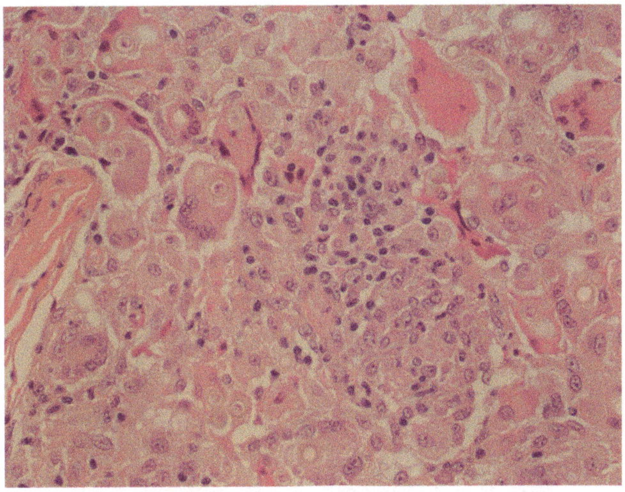

Fig. 4.21 Jorge Lobo's disease: skin biopsy, hematoxylin–eosin staining

lesions is characteristic of this infection (Fig. 4.25). Usually the injury is unilateral, preferentially in the lower limbs (foot or leg). A bacterial infection can occur over the lesion, and there is a nasty odor. Evolution into squamous-cell carcinoma can occur.

Direct mycological examination of skin scales may show typical fungal structures in an optical microscope (Fig. 4.26). These skin scales should come from the areas around the black dots of the skin lesions. Fungal culture will define the species of the etiological agent. Histopathological examination of a skin biopsy is characteristic but does not define the species of the causative agent.

Treatment is difficult, especially in advanced stages. If lesions are small, surgery can be curative. Systemic antifungals (itraconazole, ketoconazole, fluconazole, terbinafine, 5-flucytosine) can be associated with surgical treatments (e.g., excision of lesions, cryosurgical with liquid nitrogen, electrocoagulation).

4.1.7 Mycetoma

Mycetomas are a group of infectious diseases that affect the skin and subcutaneous tissue. They can be caused by various bacterial species (endogenous and exogenous actinomycetomas) and fungi (eumycetoma). They share common clinical

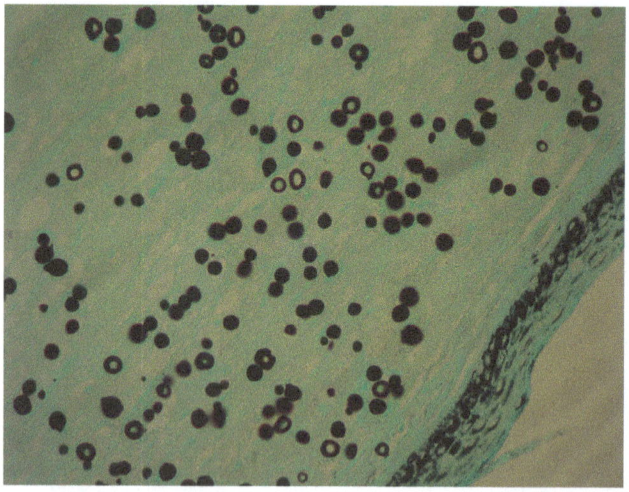

Fig. 4.22 Jorge Lobo's disease: skin biopsy, silver-methenamine staining

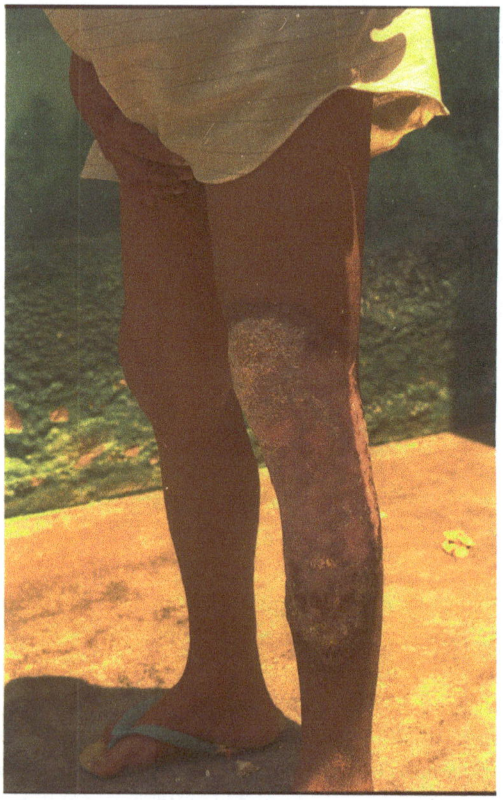

Fig. 4.23 Chromoblastomycosis: verrucous plaque

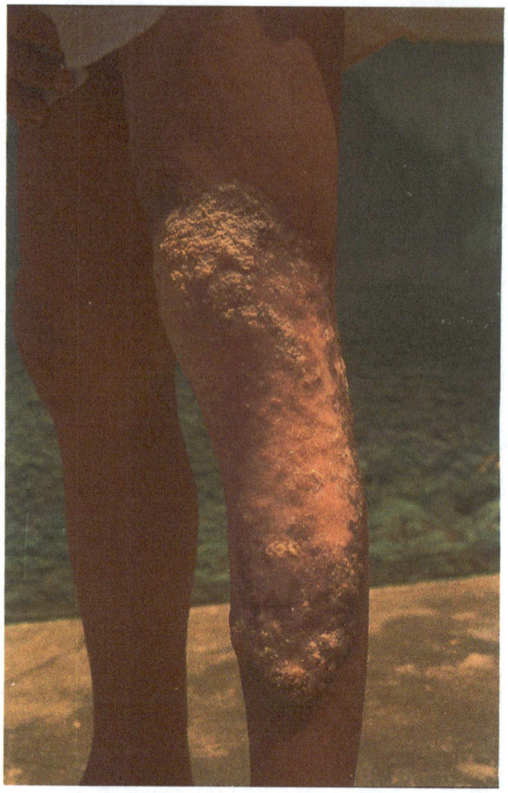

Fig. 4.24 Chromoblastomycosis: verrucous plaque

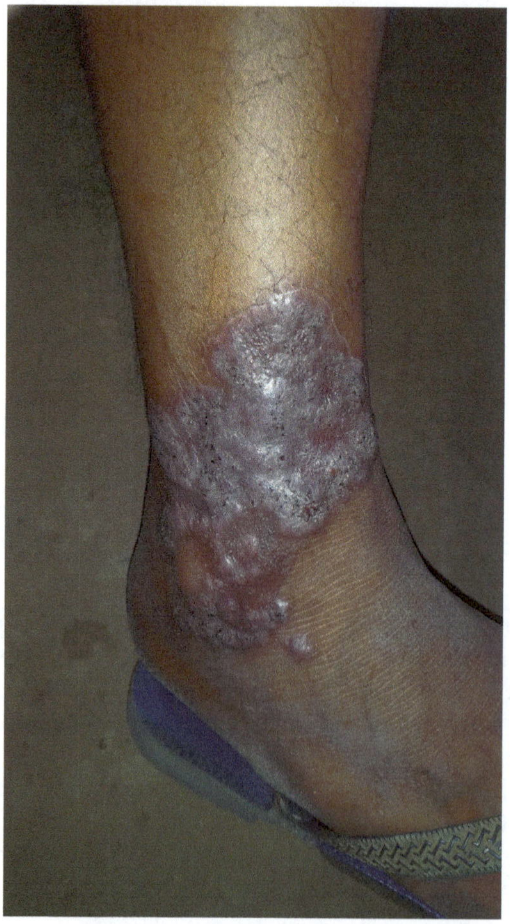

Fig. 4.25 Chromoblastomycosis: verrucous plaque with black dots

and histopathological characteristics, even if caused by different etiological agents.

Exogenous actinomycetoma and eumycetoma are more common in warm climates (tropical, subtropical, and desert areas) and are associated to people who have contact with soil and plants (rural). The disease generally affects the extremities, especially the feet. Endogenous actinomycetoma lacks this epidemiological link. Therefore, it may be found anywhere in the world and may occur following surgery.

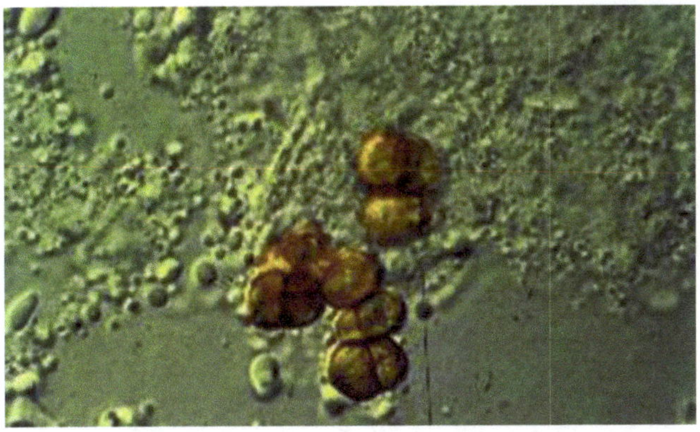

Fig. 4.26 Chromoblastomycosis: direct mycological examination of scales (100×)

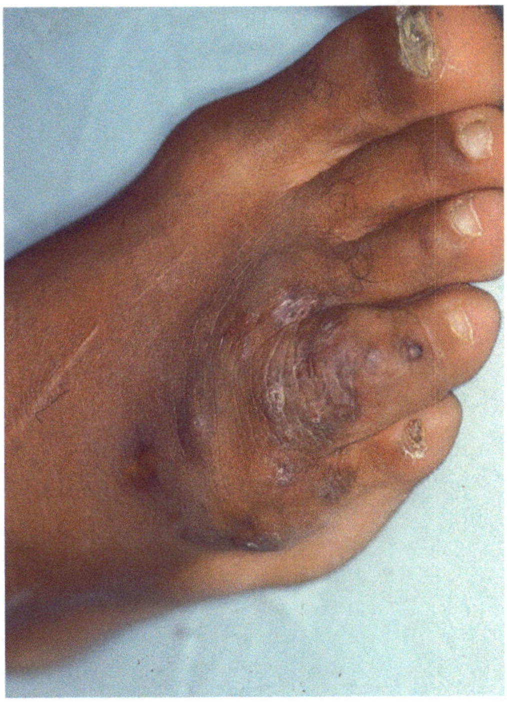

Fig. 4.27 Mycetoma: tumor lesion with fistulae

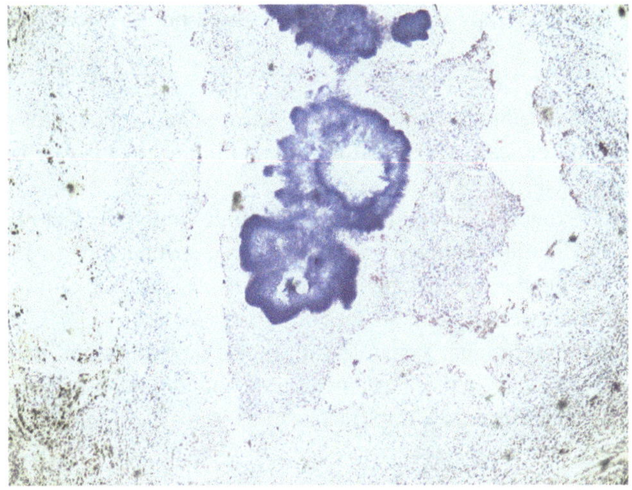

Fig. 4.28 Actinomycetoma: direct mycological examination of secretion (100×) (Photo by Orion Sant Ana Motter Borba)

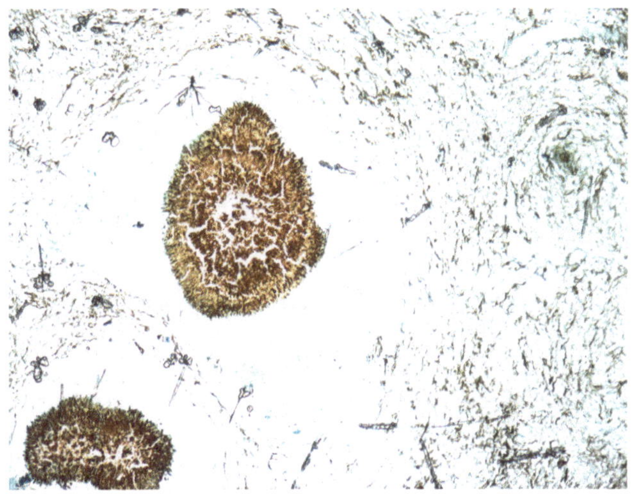

Fig. 4.29 Eumycetoma: direct mycological examination of secretion (100×) (Photo by Orion Sant Ana Motter Borba)

There are clinical and pathological similarities among the mycetomas despite the different etiological agents. Exogenous actinomycosis and mycetoma are characterized by hardened tumor lesions, microabscesses, and fistulas, in addition to secretions. In such secretions, granules and grains may be present. The lesions are usually unilateral and the lower limbs are the preferential locations (Fig. 4.27). These areas are more subject to trauma. The trauma is the likely form of inoculation of the causative agent. There is usually no pain or systemic symptoms.

Exogenous actinomycosis is caused by aerobic bacteria of different species (e.g., *Nocardia brasiliensis*), while the eumycetoma is caused by several species of fungi (e.g., *Madurella mycetomatis*). Endogenous actinomycosis is characterized by fistula-draining secretions with beads or grains, which are located in the neck, face, thorax, or abdomen. They are caused by anaerobic bacteria localized in the oral cavity, lungs, and intestines. Endogenous actinomycosis is a systemic disease on which more research is needed in other body organs.

Diagnosis is confirmed by direct examination of removed secretion through cutaneous fistulae (with the addition of 10% potassium hydroxide solution, the grains can be visualized in this material; see Figs. 4.28 and 4.29).

Histopathological examination or cultures for fungi and aerobic and anaerobic bacteria can be performed. The diagnosis of mycetoma cannot be excluded if the characteristic grains are not evidenced in the histopathological examination since often they are deep in the lesions, requiring a deep exploration of the fistulae to find them.

Drug therapy will be conducted based on the etiological agent:

(a) Endogenous actinomycosis: systemic antibiotic therapy (G-crystalline penicillin, ampicillin, tetracycline, or erythromycin);
(b) Exogenous actinomycosis: systemic antibiotic therapy (diamino-diphenyl sulfone, sulfamethoxazole-trimethoprim, rifampicin, aminoglycosides, quinolones, amoxicillin, tetracyclines or carbepenens);
(c) Eumycetoma: systemic antifungals (ketoconazole, itraconazole, terbinafine, potassium iodide, amphotericin-B, voriconazole or posaconazole).

Surgical procedures may be necessary depending on each clinical situation. For instance, if lesions are beginning and small, their complete excision can control or cure the disease.

4.2 Bacterial Diseases

4.2.1 Impetigo

Impetigo is a skin disease caused by bacteria, mainly *Staphylococcus aureus* and *Streptococcus pyogenes*. It is a common and very contagious disease that is more prevalent in children. This superficial infection may result from contamination of any preexisting injury or through the action of scratching.

In some indigenous areas, during the dry season there is a great proliferation of mosquitoes. The itching caused by the bite of these insects can evolve into abrasions that become a gateway to bacteria, often resulting in impetigo.

The disease is characterized by pustules and blisters with purulent contents and erythema at the base of the lesion. Pus and crusts may develop anywhere on the skin, but they are more common on the face and upper and lower limbs (Figs. 4.30, 4.31, 4.32, and 4.33). If not treated properly it can progress to deeper infections like erysipela.

In most cases diagnosis is made by clinical observations. If necessary, the material (pus or secretion) is collected for bacterioscopic tests and culture for bacteria. The ideal is to collect the material from a pustule. After local cleansing, break the pustule with a small sterile needle and collect the pus with a sterile swab.

Treatment consists in cleaning the lesions with water and soap or with saline solution. Topical antibiotics can be used if few cutaneous lesions (neomycin, gentamicin, 2% mupirocin, 2% fusidic acid, or 1% retapamulin, bid). If there are many lesions or if fever or other symptoms are present, use systemic antibiotics (penicillins, cephalosporins, macrolides).

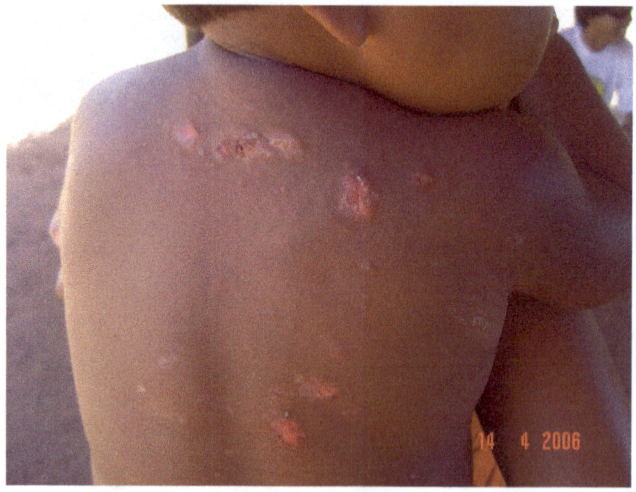

Fig. 4.30 Impetigo: skin lesions on dorsum

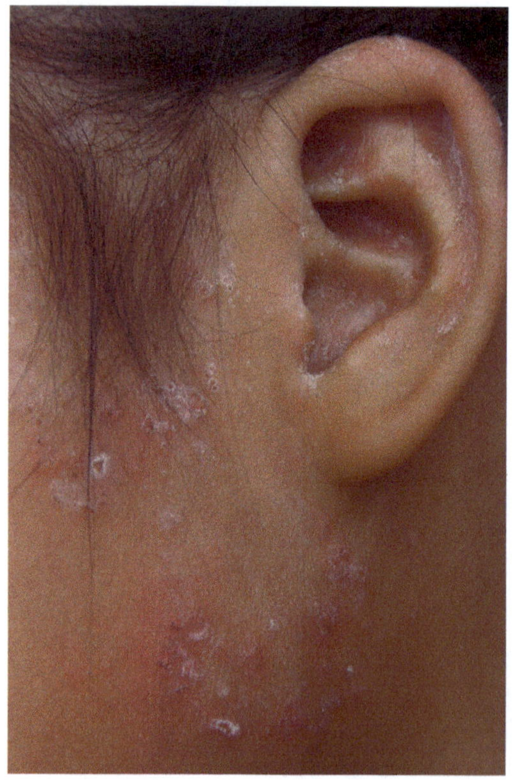

Fig. 4.31 Impetigo: skin lesions on face and ear

4.2.2 Bacterial Folliculitis

Bacterial folliculitis is a skin infection caused by *Staphylococcus aureus*. It is related to hair follicles and can be superficial or deep. It can occur at any age. The deeper forms occur in adults. Bacterial folliculitis does not appear to be more frequent in indigenous compared to Caucasian or African populations.

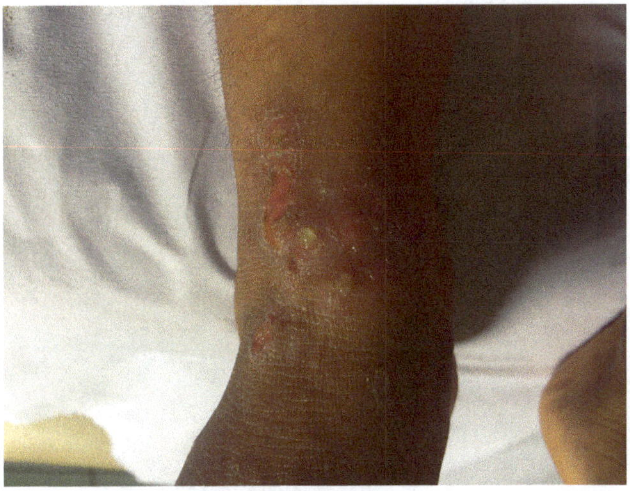

Fig. 4.32 Impetigo: skin lesions on leg

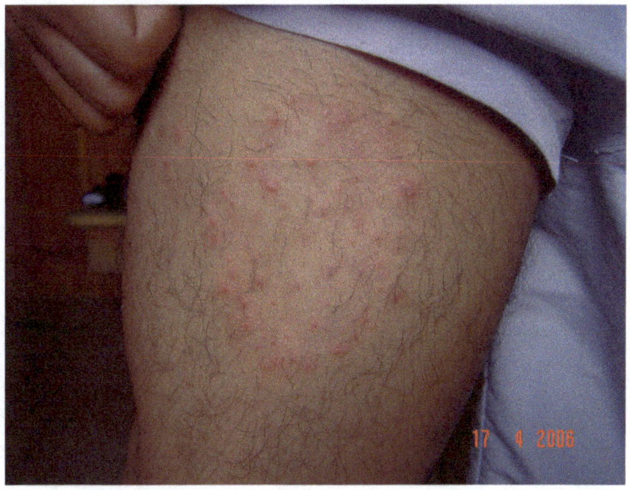

Fig. 4.34 Bacterial folliculitis: skin lesions on thigh

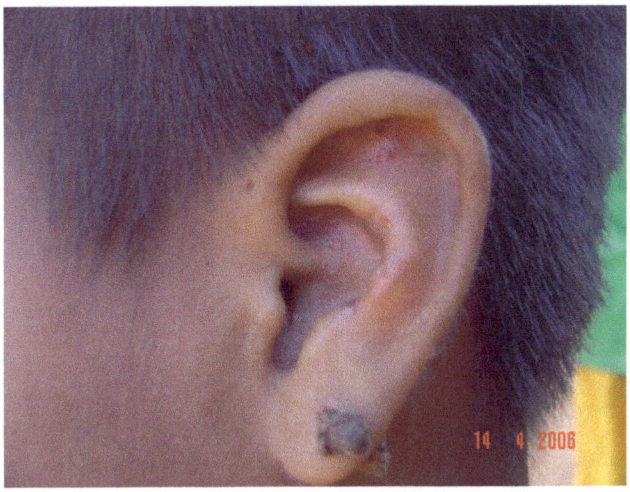

Fig. 4.33 Impetigo: skin lesions on ear

The superficial form is also known as ostiofollulitis or impetigo of Bockhart and presents as a pustule in the hair follicles. When breaking, it evolves with the formation of crust. It does not destroy hair follicles. It is a form of impetigo of follicular localization (Fig. 4.34).

The deep form, called *sycosis barbae*, is characterized by follicular pustules that can progress to a chronic state if not properly treated. It should be differentiated from *trichophytic sycosis* (caused by fungi). Mycological and bacterioscopic examinations are necessary for differentiation, since their clinical pictures may be indistinguishable.

Another deep form, the hordeolum, affects the eyelashes (meibomian glands) with the formation of pustules or painful erythematous nodules. Pruritic lesions on the eyelid, seborrheic dermatitis on the eyelashes, and refractive defects may be predisposing factors. Fungi or bacteria cultures are indicated when there is doubt about the etiological agent. Antibiotic therapy can be used if fungal etiology is ruled out. The treatment is similar to impetigo and consists in cleaning the lesions with water and soap or with saline solution. In extensive lesions, systemic antibiotics are used (penicillins, cephalosporins, macrolides).

4.2.3 Erysipelas and Cellulitis

Erysipelas and cellulitis are skin and subcutaneous tissue diseases caused by bacteria, mainly beta-hemolytic streptococci. They are common diseases and can occur at any age. They can arise from complications related to other skin diseases such as bacterial diseases (e.g., impetigo), fungal diseases (such as of the foot), or skin ulcers (traumatic or varicose ulcers). Fever, chills, and a bad general condition are frequent symptoms in afflicted patients.

The affected area of the skin becomes reddish (erythematous), edematous, hot, and painful (Fig. 4.35). Blisters with purulent contents can occur, representing a sign of disease severity (*Staphylococcus aureus* infection). Face and lower limbs are generally more affected, but any area of the body can be involved. Cellulitis occurs when the infectious condition is deeper. If it occurs on the face, it is quite serious. In patients with diabetes mellitus, obesity, or varicose veins in the legs, erysipelas and cellulits are more common.

The etiological agent usually penetrates through some damaged part of skin, for example, foot mycoses, ulcers, or impetigo (a so-called gateway). Complications are not more frequent among indigenous people compared to populations living in urban environments.

Diagnosis is based on patient history and physical examination. Laboratory exams are not necessary in most cases. For treatment, the use of a systemic antibiotic is mandatory

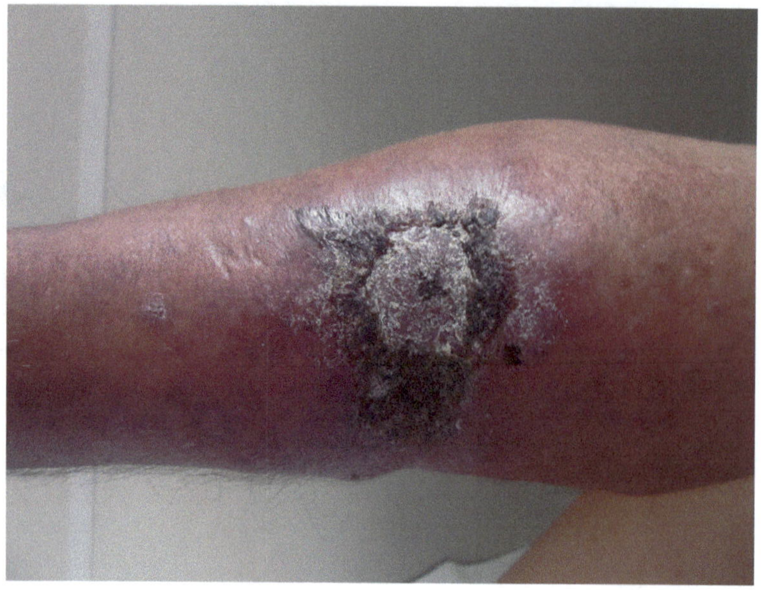

Fig. 4.35 Erysipelas erythema, edema, and crusts

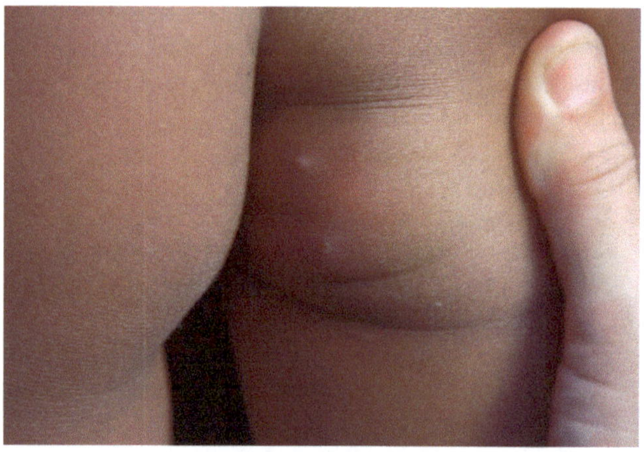

Fig. 4.36 Furuncles on buttocks

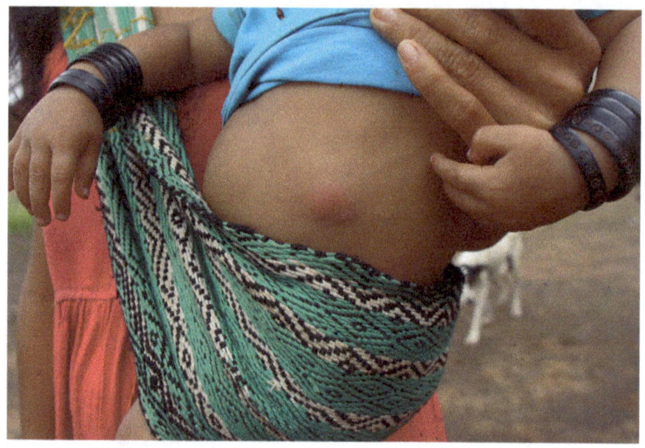

Fig. 4.37 Abscess on abdomen

(penicillin, cephalosporin, macrolide, sulfa). Any predisposing conditions or any cutaneous disease (the gateway) present should be treated.

4.2.4 Furunculosis and Abscess

Furuncles affect the pilosebaceous complex. In general, the infectious agent is a *Staphylococcus* species. They are more common in adults and may result from ingrown hairs. It is a process that can result in a scar. A furuncle begins as a follicular pustule or erythematous nodule with increased local temperature and pain. It becomes liquefied and drains, giving rise to pus and elimination of the "carnage." Legs, thighs, and buttocks are the most commonly affected sites; however, any part of the body can be affected (Figs. 4.36 and 4.37). Anthrax is formed by several confluent furuncles. By autocontagion they form painful and hot erythematous nodules, in addition to points where the pus drains out. A large collection of pus is called an abscess.

Furunculoid myiasis may be confused with furuncles. The presence of a larval breathing orifice and the exit of persistent serous secretions aid this differentiation. The patient reports a sensation of intermittent pain at the site, while in the furuncle the pain is more continuous.

Culture for bacteria is indicated when there is diagnostic doubt. The material can be collected with sterile cotton and placed in an appropriate culture medium. Treatment is similar to that for impetigo, always using systemic antibiotics (penicillin, cephalosporin, macrolide). In cases of resistance to treatment, Community-associated methicillin-resistant *Staphylococcus aureus* (CA-MRSA) consider.

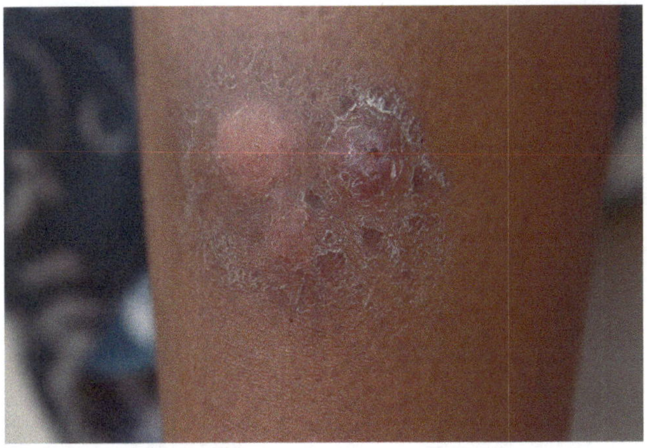

Fig. 4.38 Mycobacteriosis on leg

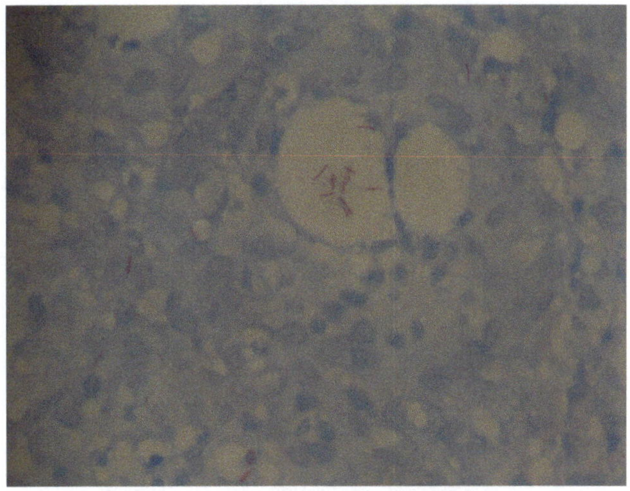

Fig. 4.39 Mycobacteria: skin biopsy, Ziehl–Neelsen stain

4.2.5 Cutaneous Mycobacteriosis

The mycobacteria diseases known as mycobacteria other than tuberculosis (MOTT) or nontuberculous mycobacteria (NTM) can occur in both immunocompetent and immunocompromised individuals. They are caused by mycobacteria, except for the complex *Mycobacterium tuberculosis* (tuberculosis) and *Mycobacterium leprae* (leprosy). The reservoir of these mycobacteria is broad. They can be found in fish, poultry, cattle, monkeys, soil, water, and humid places. Dozens of MOTTs have been described. About one-third can cause disease in humans.

Clinical presentation varies according to the immune status of the host. In immunocompromised hosts it may appear as localized or disseminated disease, accompanied by general symptoms such as fever and asthenia. The lung may be the initial focus. Generally localized skin lesions are inflammatory, papular or nodular, and painless and do not evolve beyond the site of agent inoculation (Fig. 4.38).

The mycobacteria can be found in the biopsy of the skin by the Ziehl–Neelsen or Fite–Faraco stain (Fig. 4.39). Identification of the agent by culture is essential to characterizing the disease. The culture may be made of purulent material aspirated from cutaneous lesions or from skin fragments collected by biopsy.

There is no consensus on the treatment of different forms of the disease. If possible (in the case of small lesions), a surgical resection associated with systemic antibiotic therapy is indicated. Depending on the etiological agent identified in culture, antibiotic therapy is performed (clarithromycin, azithromycin, ciprofloxacin, ofloxacin, levofloxacin, doxycycline, minocycline, imipenem, amikacin, rifampicin, ethambutol, pyrazinamide, isoniazid, streptomycin). Combination therapy (association of two drugs) is recommended. Treatment regimens can last from 1 to 6 months on average.

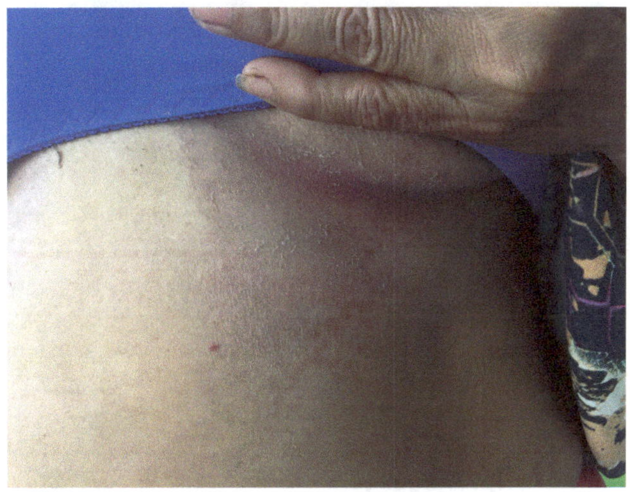

Fig. 4.40 Erythrasma: lesions on folds of skin

4.2.6 Erythrasma

Erythrasma is a chronic bacterial infection caused by *Corynebacterium minutissimum*. It affects mainly folds of the skin (interdigital regions of the toes, armpits, inguinal region) and is related to increased sweating. Obesity and diabetes mellitus are conditions that favor the onset of the disease. It affects men and women in equal proportions.

Diagnosis is clinical. Skin lesions are well-defined scaly spots or plaques with irregular forms, brown or erythematous-brownish colored (Figs. 4.40 and 4.41). In Wood's light cutaneous lesions show the typical red-coral color (fluorescence due to the production of coproporphyrin III by the bacterium). The diagnosis can be confirmed by collecting skin scrapings for direct bacterioscopic examination by Gram staining (Gram-positive coccobacilli). Differential diagnoses are dermatophytosis, candidiasis, and psoriasis.

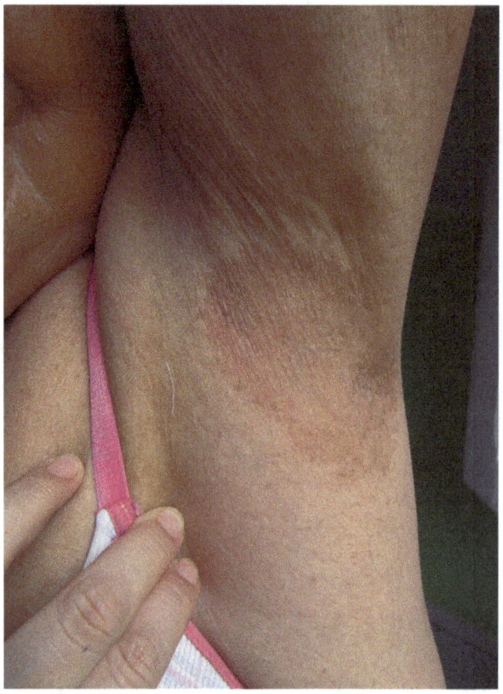

Fig. 4.41 Erythrasma: lesions on armpit

Treatment involves improving predisposing local conditions (sweating, humidity). Topical 2% erythromycin or imidazole antifungal medications are used twice daily until complete improvement is achieved.

Oral erythromycin can be used in cases involving extensive lesions or therapeutic failure with topical medications.

4.3 Viral Diseases

4.3.1 Warts

Warts are a skin disease caused by human papillomavirus virus (HPV). It is a common disease and frequent in children and adults. Transmission occurs through personal contact. Skin lesions are typical. The main characteristic is the verrucous surface. Warts can vary in size, be single or multiple, and affect different parts of the body. When a lesion is single or small, with a rather verrucous and irregular surface, it is called *verruca vulgaris* (Fig. 4.42). Warts in the feet can be painful when walking. Lesions may be located in the cuticle region (periungual warts). When a lesion is single or multiple with flattened and pinkish papules, it is called *verruca plana*; it occurs commonly on the hands, feet, elbows, and knees (Figs. 4.43, 4.44, and 4.45). The Koebner phenomenon is the appearance of new lesions in areas of the skin that have received minor trauma (Fig. 4.46).

Some types of HPV can involve transformation of lesions into cancer. Genital warts in women can result in cancer of the cervix, so diagnosis and treatment are very important.

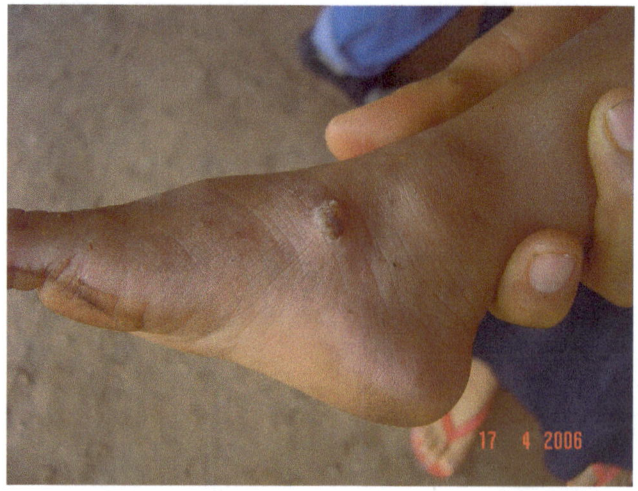

Fig. 4.42 Warts: verruca vulgaris on foot

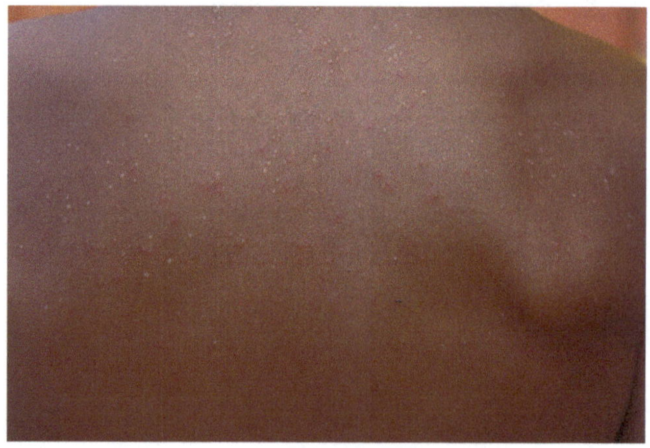

Fig. 4.43 Warts: verruca plana on dorsum

The diagnosis of viral infection by HPV is clinical. If there is diagnostic doubt, perform a skin biopsy for anatomopathological examination.

Treatment can include topical keratolytic drugs like salicylic acid or lactic acid, which should be applied carefully to warts once daily. Imiquimod cream, 5%, is another topical option in some cases. Other procedures can also be used by health professionals, such as chemical surgery with concentrated nitric acid or 80% trichloroacetic acid. Other possible methods, like surgical therapy, electrocautery, cryosurgery (with liquid nitrogen), laser therapy, and photodynamic therapy, are specialized and require special equipment and materials. Infection by some oncogenic HPVs can be prevented by vaccination.

4.3.2 Focal Epithelial Hyperplasia

Focal epithelial hyperplasia is a benign and contagious disease caused by HPV, especially HPV-13 and HPV-32 types. It is a rare condition that affects some countries and some

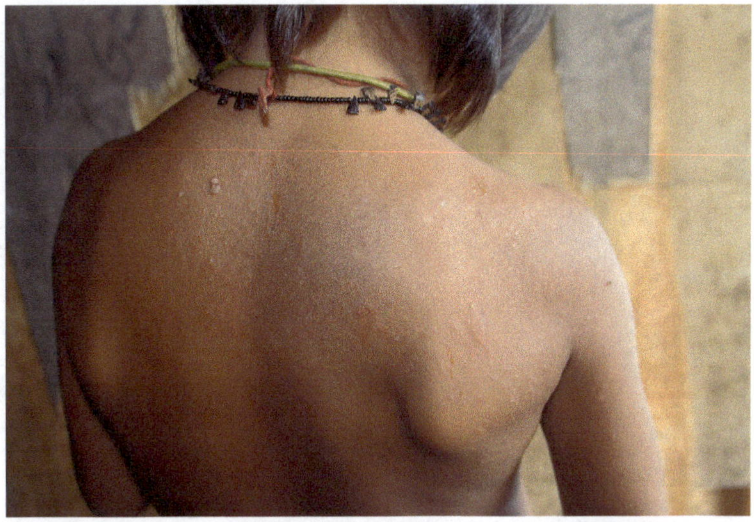

Fig. 4.44 Warts on dorsum

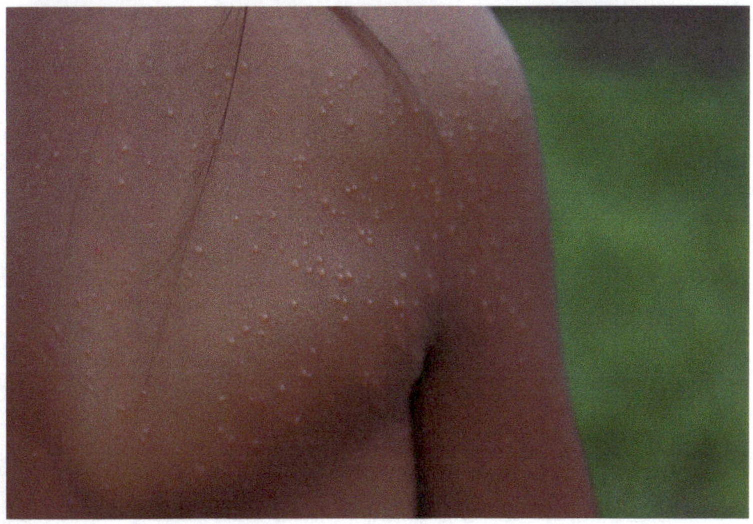

Fig. 4.45 Warts: verruca plana on the dorsum and arm

people, including some indigenous ethnic groups. There may be cases in the same family and mostly among children. Familial cases of this disease have been observed in the Xingu Indigenous Park, Brazil. It is characterized by small asymptomatic verrucous papules affecting mainly oral mucosa, tongue, and lips (Figs. 4.47, 4.48, and 4.49). Eventually there may be injuries other areas of the body.

Diagnosis is clinical. A skin biopsy may be performed and the material submitted for anatomopathological examination. Usually no treatment is necessary because the lesions tend to heal spontaneously. When treatment is needed, the main options are chemical surgery with trichloroacetic acid, electrocautery, and cryosurgery.

4.3.3 Condyloma Acuminatum

Condyloma acuminatum is a disease caused by HPV localized in the genital skin and mucosa. It is a sexually transmitted disease (STD) that is highly transmissible and difficult to eliminate. Some types of virus can cause carcinomas in both the cervix and the penis. Skin lesions are similar to those of the verruca vulgaris with genital location (Fig. 4.50). In men they are located in the penis and can occur in large numbers and attain large sizes (appearance of cauliflower). In women they are located in the vulva and vagina but can be internal and not noticed by the patient.

For therapy, medications that slowly remove warts, such as podophyllin colodium 25%, are used. The substance should be carefully spread over the warts and subsequently removed (by washing) after 4 h. Application is done once every 3 days until the lesions disappear. Other medications, such as 50% trichloroacetic acid, 0.5% podophyllotoxin cream, or 5% imiquimod cream, may also be used.

Procedures such as surgical therapy, electrocautery, cryosurgery, laser therapy, and photodynamic therapy can be performed, but these rely on specialized equipment.

Partners should always be examined and treated if necessary. Patients are advised to always use condoms during sexual intercourse since the virus is transmissible and difficult to cure. Infection by certain oncogenic HPVs can be prevented by vaccination.

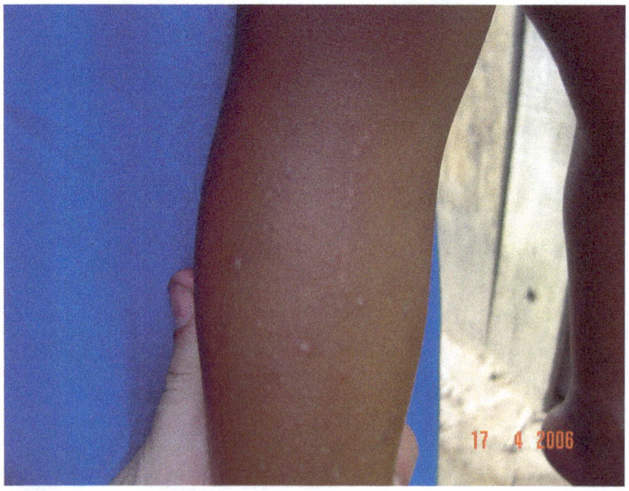

Fig. 4.46 Warts: Koebner phenomenon

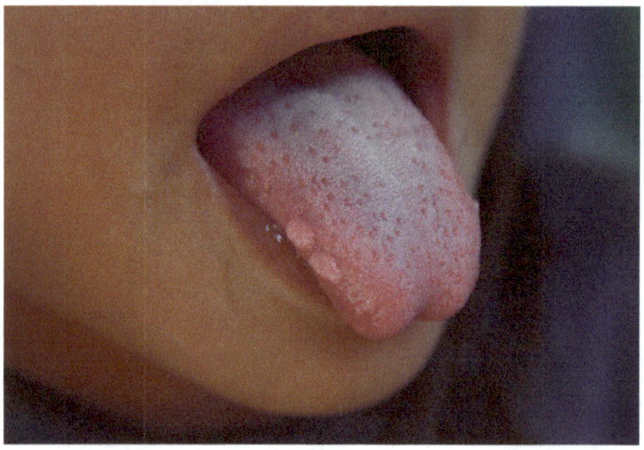

Fig. 4.47 Focal epithelial hyperplasia: lesions on tongue

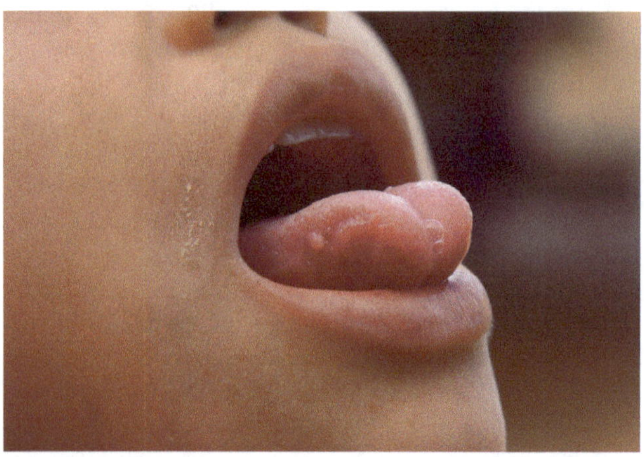

Fig. 4.48 Focal epithelial hyperplasia: lesions on tongue

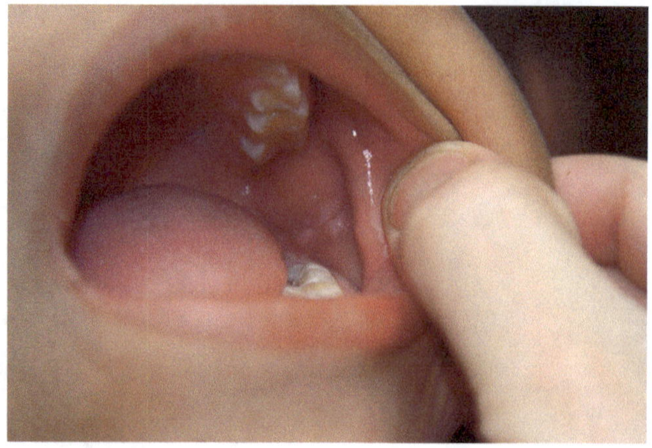

Fig. 4.49 Focal epithelial hyperplasia: lesions on mouth

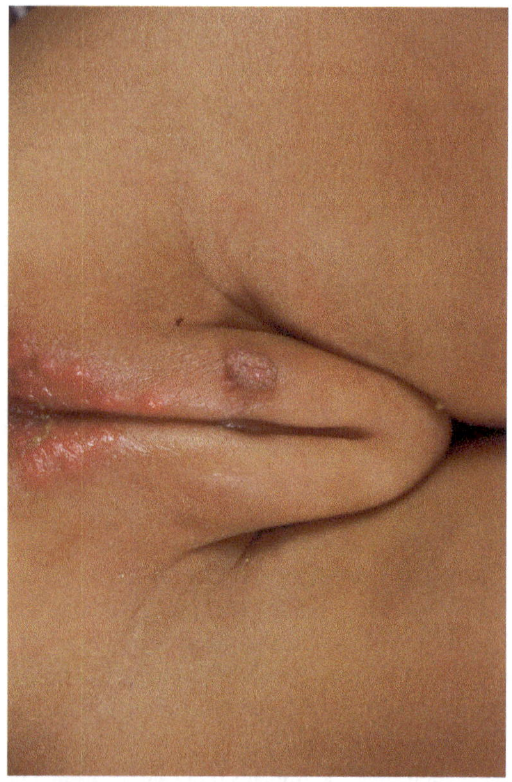

Fig. 4.50 Condyloma acuminatum: lesions on vulva

4.3.4 Molluscum Contagiosum

Molluscum contagiosum is a skin disease caused by the poxvirus group. It occurs more frequently in childhood, after lactation, and before puberty. Transmission occurs through personal contact, and it is very contagious among children. It is rare in adults but may occur in pregnant women and can be transmitted sexually. It is not a serious illness, and often

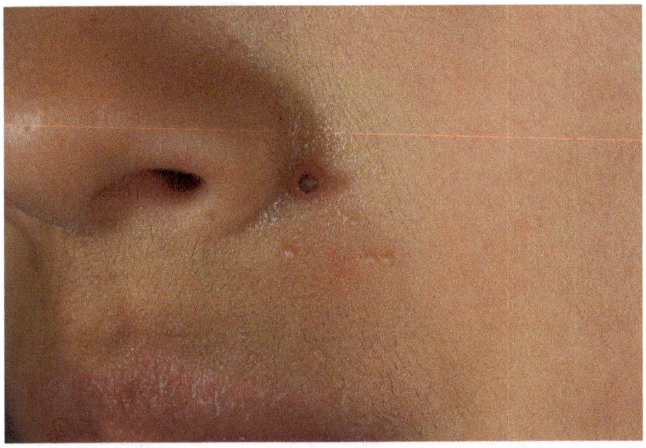

Fig. 4.51 Molluscum contagiosum: lesions on face

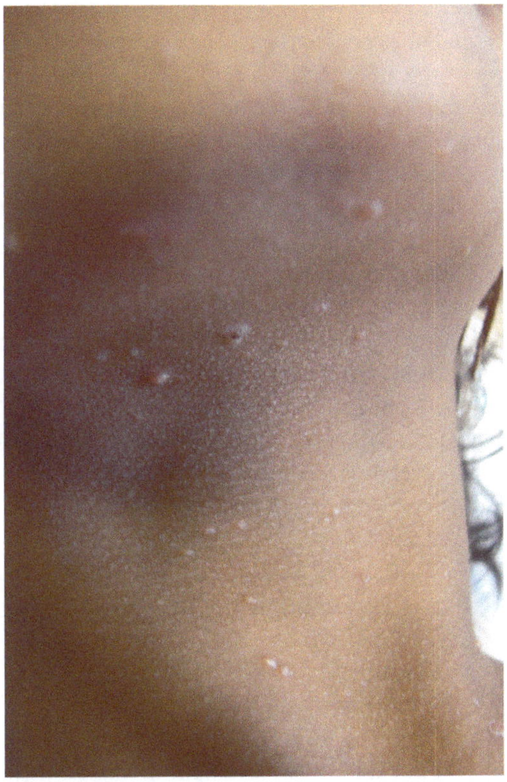

Fig. 4.52 Molluscum contagiosum: lesions on neck

healing occurs spontaneously after 1 year, but it causes great discomfort. Skin lesions from this disease are different from common wart lesions. They are small and firm and have a shiny, smooth surface. Usually there is a depression in the middle of the lesions (central umbilication) (Figs. 4.51 and 4.52). Large numbers of lesions are common. Any area of the body can be affected, but they occur more frequently in warmer areas such as folds of the armpit, elbow, knees, neck, and genitals. Lesions are asymptomatic. Eczema may occur

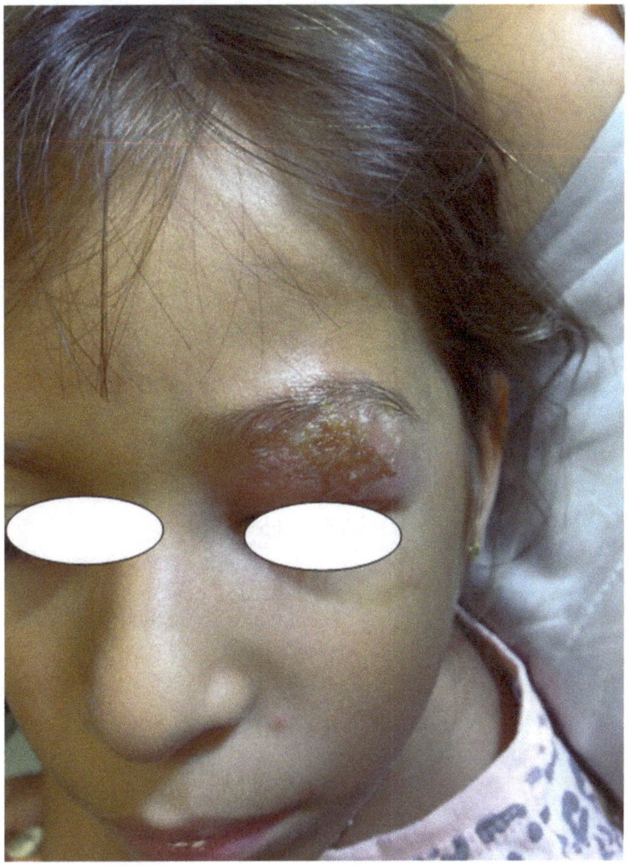

Fig. 4.53 Herpes simplex: lesions on face

around the lesions, leading to desquamation, erythema, and local itching in children with allergies.

Diagnosis is essentially clinical. Treatment can involve salicylic acid in an elastic collodion, which should be applied carefully on the lesions once daily. Other procedures are also possible, such as curettage of the lesions, which is an easy procedure but requires special training and material (a curette). Cryotherapy or the application of 50% trichloroacetic acid can also be carried out.

4.3.5 Herpes Simplex

Herpes simplex is a skin disease caused by the human herpesvirus (HHV). The affected individual comes in contact with this virus in childhood, and the virus becomes latent in the nervous ganglion. It manifests when the individual is immunocompromised by stress, excessive exposure to the sun, fever, or infection.

It is a virus of universal distribution, which can occur between children and adults. The most commonly affected site is the face (Fig. 4.53), especially on or near the lips (Fig. 4.54). The genital mucosa may also be affected. In general the virus involved is

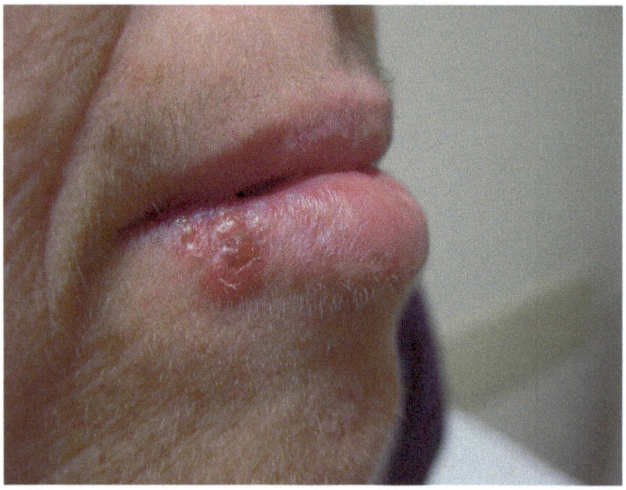

Fig. 4.54 Herpes simplex: lesions on lips

HSV-2. It occurs in patients who are already sexually active and can be a risk factor for other infections. It is characterized by the presence of translucent vesicles on a erythematous-edematous base arranged in a grouped distribution (Fig. 79). The lesions are located mainly in the lips, perioral region, or in the genital region (penis or vulva). Symptoms range from tingling, itching, and burning at the site of the lesions.

Diagnosis is essentially clinical. One can conduct a cytological test by smearing the floor of a vesicle. Local treatment involves local hygiene and antiviral cream (acyclovir, penciclovir). If the disease is very intense, systemic antivirals (acyclovir, valacyclovir, famciclovir) are indicated.

4.3.6 Herpes Zoster

Herpes zoster is a skin disease caused by the varicella-zoster virus. This virus causes chickenpox during childhood and can reoccur years later in the form of herpes zoster. Following the initial infection, the virus may become latent in the nervous ganglion; however, in an immunosuppressive state, it multiplies and spreads to the skin. The disease can be a manifestation of immunodepression status caused by human immunodeficiency virus (HIV) infection, systemic neoplasms, or iatrogenesis (use of immunosuppressive drugs such as systemic steroids).

Chickenpox is a common disease in childhood. Currently, the vaccine prevents the first infection. Herpes zoster is a manifestation that depends on the state of the individual's immunity. Herpes zoster infection comes from the virus; however, its transmissibility is lower than that of chickenpox.

Herpes zoster is more frequent in adults or in the elderly. It is characterized by the presence of vesicles and bubbles with a translucent, hemorrhagic (bloody), or purulent (yellowish) content on an erythematous and edematous basis.

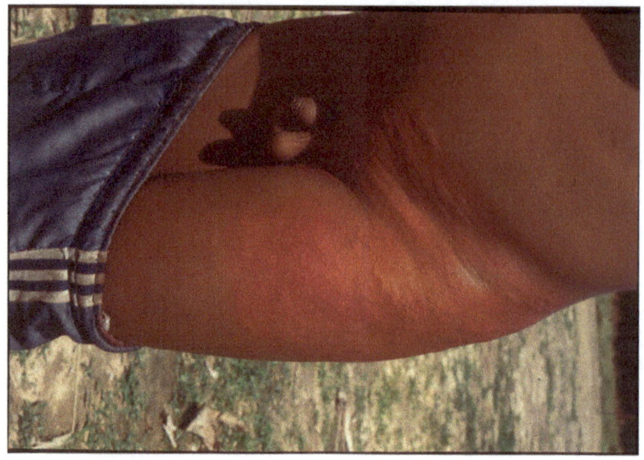

Fig. 4.55 Herpes zoster: lesions in linear distribution

Vesicles and blisters are arranged linearly following the distribution of the affected nerve (Fig. 4.55). It occurs only on one side of the body and is accompanied by pain or a tingling sensation. When it occurs on the face, treatment should be redoubled because it can affect the cornea of the eye, leading to corneal ulcers and loss of vision.

Diagnosis is essentially clinical. A smear may be collected from the blister floor (scraped from the base of the bubble after rupture) and painted to examine the cells. This cytologic test shows the changes that characterize herpes virus infection. Systemic treatment with antivirals is the most indicated therapy (acyclovir, famcyclovir, or valacyclovir).

If the patient experiences pain at the injury site, painkillers are indicated. It is always important to consider the possibility of postherpetic neuralgia.

4.4 Parasitic Diseases

4.4.1 Scabies

Scabies is a skin disease caused by the parasite *Sarcoptes scabiei*. It is a common disease, more frequent in children, very contagious, and can cause epidemics, such as the one that occurred among the Panará Indians when they inhabited the Xingu Indigenous Park. Transmission occurs through personal contact and contaminated clothing (e.g., shirts, nets, towels, shorts). Scabies occurs in domestic animals, such as dogs and cats. In general, affected animals exhibit hair loss. The animal disease can eventually be transmitted to humans, especially those in close contact with infected animals.

Scabies causes a lot of itching, especially at night when the patient is lying down. Characteristic lesions are erythematous papules. Small wounds with scabs or abrasions appear in skin folds (armpits, between the fingers, groin) but may invade other areas (pelvic girdle, wrists, umbilical region) and eventually the entire body (Figs. 4.56, 4.57, and 4.58). In women, it commonly affects the breasts. In men, it is common for lumps to appear in the penis and scrotal sac (Fig. 4.59). Children can develop blisters on the palms of the hands and soles of the feet. Usually more than one person in the same home or family will experience itching. Bacterial infection in lesions is common.

In almost all cases, diagnosis is made by clinical presentation and epidemiology (family cases). Usually treatment with topical medications is sufficient (permethrin, sulfur precipitate, benzyl benzoato). If necessary, systemic drugs (ivermectin repeated after 1 week) can be used. For treatment success, relatives and those who live in the same house must be treated. All clothes should be washed and dried daily.

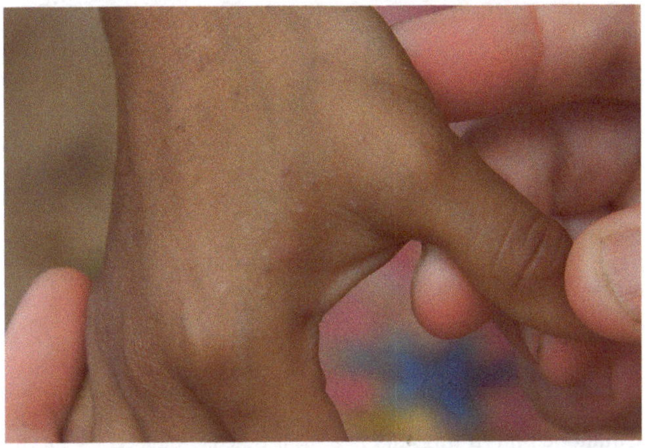

Fig. 4.56 Scabies: lesions between fingers

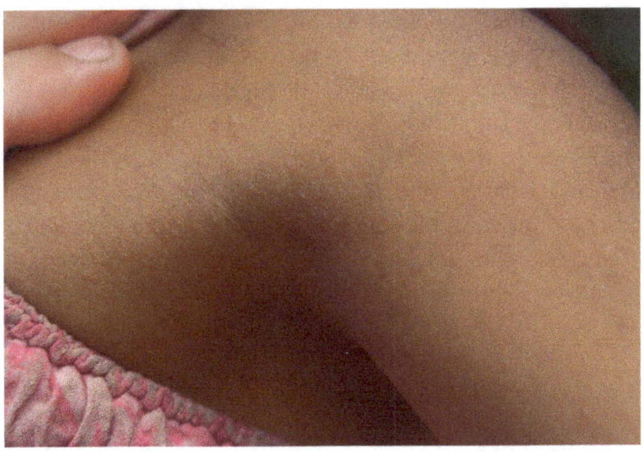

Fig. 4.57 Scabies: lesions on armpit

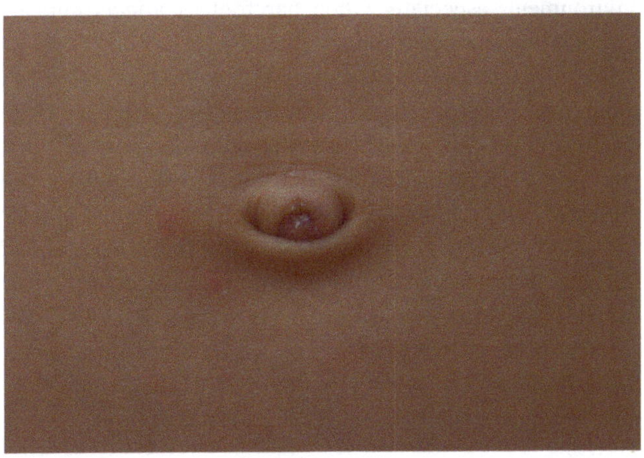

Fig. 4.58 Scabies: lesions in umbilical region

4.4.2 Pediculosis

Pediculosis is a disease of the hair caused by parasites of the genus *Pediculus* (lice). When it reaches the head it is called *pediculosis capitis*. When it reaches body hairs it is known as *pediculosis corporis*. When it affects pubic hair, it is called *pediulosis pubis* or *phthiriasis*.

It is a common disease, especially among children, very contagious, and can cause epidemics. Transmission is through personal contact. Phthiriasis can also be transmitted through sexual contact. It causes a lot of itching, especially on the head, the most frequently affected site. Abrasions can occur on the back just below the nape of the neck. This form usually begins in the posterior (occipital) region.

Diagnosis is made by visualizing the lice in the affected area and by the presence of the eggs of these

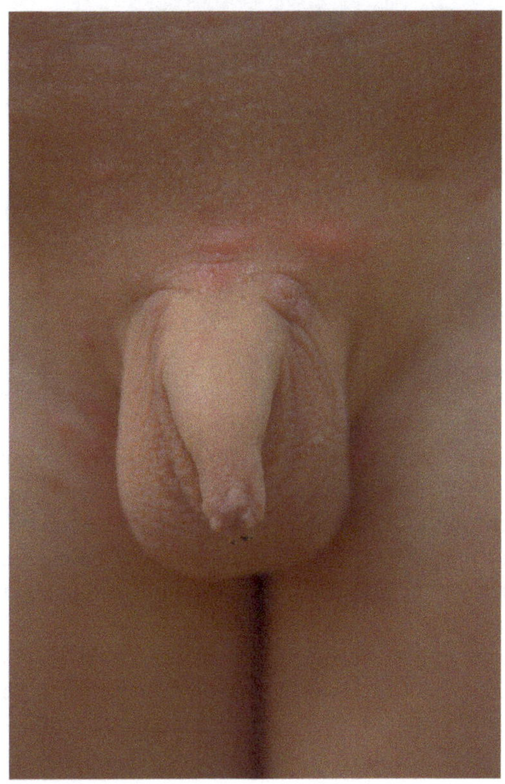

Fig. 4.59 Scabies: lesions in male genital region

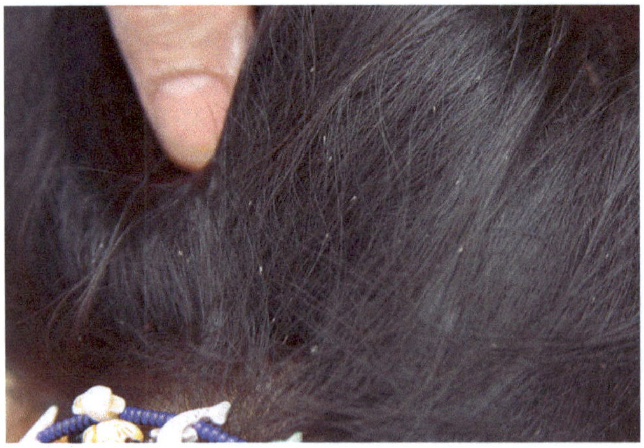

Fig. 4.60 Pediculosis capitis: nits in hair

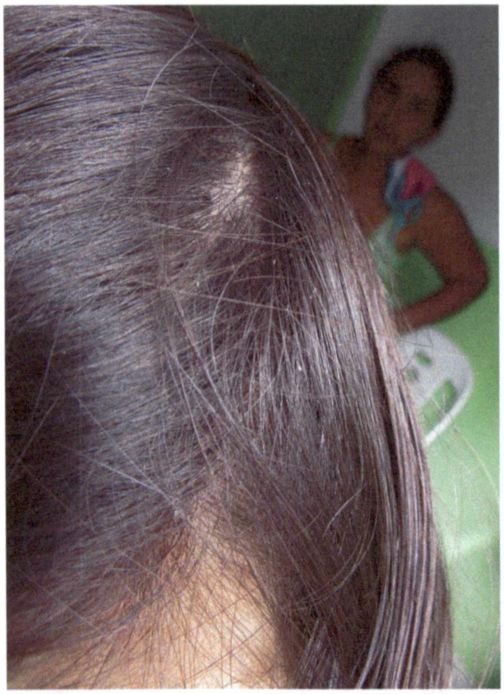

Fig. 4.61 Pediculosis capitis: nits in hair

parasites (nits), which adhere to the hair (Figs. 4.60 and 4.61). Nits can be observed clinically without examination through a microscope. Generally treatment with topical medication is sufficient (benzyl benzoate, permethrin). If necessary, systemic medications (ivermectin) can be used.

For successful treatment, it is also necessary to treat others with whom the infected person is in personal contact, especially children. Nits should be removed with a thin comb, after softening them by applying warm water on the head for several minutes.

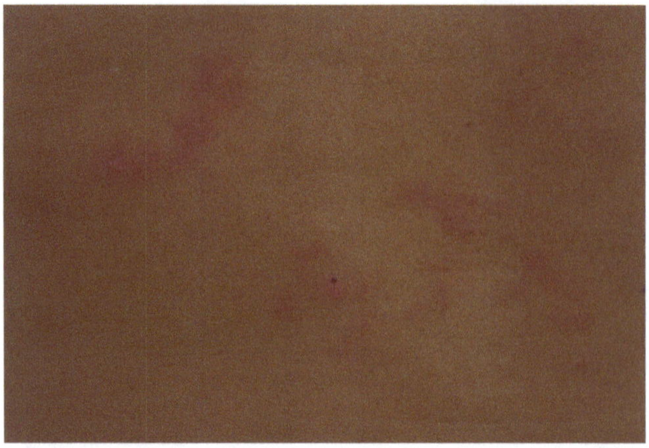

Fig. 4.62 Cutaneous larva migrans: lesions with linear and serpiginous aspect

4.4.3 Cutaneous Larva Migrans

Cutaneous larva migrans is a skin disease caused by larvae present in soils that are deposited in the feces of dogs and cats. It is a common disease in places with sandy and dry soil. It is not transmitted from person to person. It causes considerable itching in affected areas, which are usually the feet, but can also be the hands, buttocks, or any other region of the body. Cutaneous lesions are characteristic, with linear and serpiginous aspects, and increase with time by the movement of the larvae (Figs. 4.62 and 4.63). Diagnosis of cutaneous larva migrans infestation is clinical. Topical treatment is effective (thiabendazole), and the use of systemic medications (ivermectin, albendazole, thiabendazole) is an option.

4.4.4 Tungiasis

Tungiasis is a skin disease caused by a flea called *Tunga penetrans*. The disease also goes by regional names. The etiological agent is found in dry, sandy soil (land, rural areas). It penetrates the skin of a person who is in contact with this environment, especially when barefoot. It affects humans and pigs and is more common in adults.

Initially it causes slight itching in the affected region and progresses to cause local pain sensations. Characteristic lesions are yellowish papules with central dark spots, which represent the agent introduced in the skin (Fig. 4.64). The preferred location is in the foot region (interdigital plantar and subungual), but occasionally it can affect other areas of the body. Bacterial secondary infection is a frequent complication. If the infestation is massive in the feet, it may be confused with other dermatoses. In almost all cases, diagnosis is made through clinical examination and epidemiology (history of contact with soil or sand, for example).

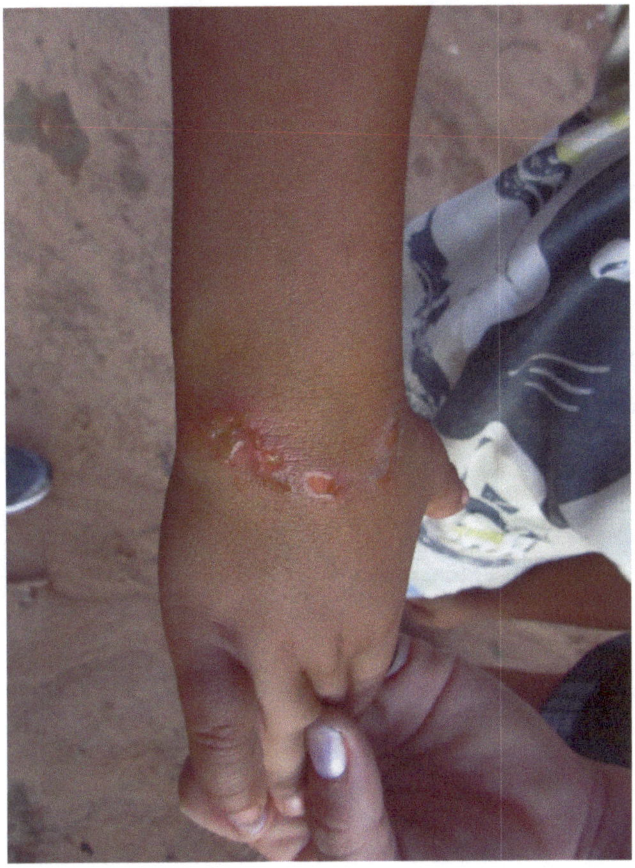

Fig. 4.63 Cutaneous larva migrans: lesions with linear and serpiginous aspect

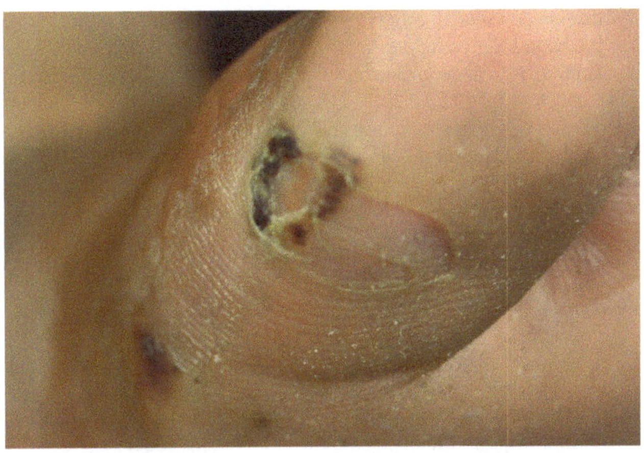

Fig. 4.64 Tungiasis: yellowish papule with central dark spot

Treatment consists of the withdrawal of the etiological agent with a sterile needle or other means (surgery, curettage, electrocoagulation, cryotherapy with liquid nitrogen). Systemic thiabendazole (3–5 days) or oral ivermectin (single dose) may be used in cases of massive infestation.

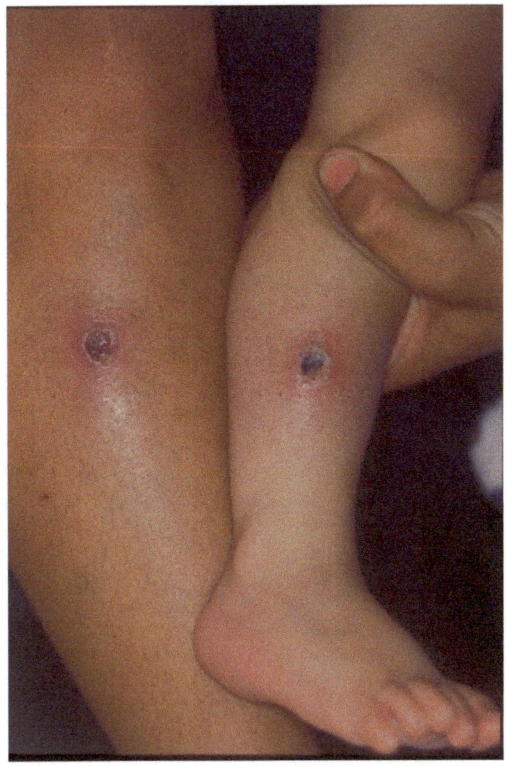

Fig. 4.65 Tegumentary leishmaniasis: characteristic ulcer; father and son with same disease

4.4.5 Tegumentary Leishmaniasis

Tegumentary leishmaniasis is a skin and mucosal disease caused by parasites of the genus *Leishmania*. It occurs forested areas and is transmitted by bites of some types of mosquitoes that inhabit these regions and that are contaminated with the parasite. The reservoirs of the disease are domestic animals and horses. Transmission does not occur directly from contact with an infected animal or human. A mosquito always plays a role in transmission.

In many regions of Brazil, the most common species is *Leishmania braziliensis*. This species is responsible for damage to the skin and mucous membranes (nose and mouth). The typical skin lesion is an ulcer with rounded contours, elevated and well-defined edges, and a reddish basis (Fig. 4.65). However, the disease can have different clinical presentations (Fig. 4.66).

Lesions are asymptomatic, do not hurt, do not itch. The disease can heal spontaneously after a few months and reappear years later, causing lesions on mucous membranes, especially of the nose. Because the parasite has a predilection for cartilage, it installs itself in nasal cartilage, destroying it and piercing it. The patient develops a deformity at the tip of the nose, with the appearance of *tapir nose*.

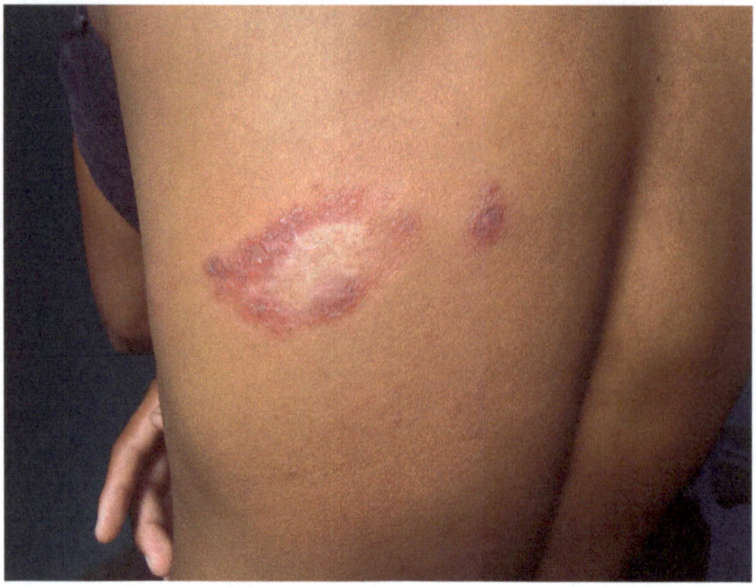

Fig. 4.66 Tegumentary leishmaniasis: plaque on dorsum

Diagnosis can be performed by a small biopsy on the edge of a lesion. The pathology shows a noncaseating granuloma. An intradermal application test, called a Montenegro test, can be conducted in which 0.1 mL of a reagent (Montenegro antigen) is injected and the papule that forms at the injection site after 48–72 h is examined. If the papule is larger than 5 mm, the likelihood that the disease was caused by *Leishmania* is strong. Treatment involves systemic medicines (meglumina, itraconazol, allopurinol). When systemic medications are contraindicated, intralesional infiltration with meglumine may be performed without dilution, using at most one ampoule (5 mL) per application.

Suggested Reading

Dermatophytosis

1. Kaushik N, Pujalte GG, Reese ST. Superficial fungal infections. Prim Care. 2015;42:501–16.
2. Van Zuuren EJ, Fedorowicz Z, El-Gohary M. Evidence-based topical treatments for tinea cruris and tinea corporis: a summary of a cochrane systematic review. Br J Dermatol. 2015;172:616–41.
3. Bressan AL, Silva RS, Fonseca JC, Alves MF. Majocchi's granuloma. An Bras Dermatol. 2011;86:797–8.

Paronychia

4. Shafritz AB, Coppage JM. Acute and chronic paronychia of the hand. J Am Acad Orthop Surg. 2014;22:165–74.
5. Relhan V, Goel K, Bansal S, Garg VK. Management of chronic paronychia. Indian J Dermatol. 2014;59:15–20.
6. Rigopoulos D, Larios G, Gregoriou S, Alevizos A. Acute and chronic paronychia. Am Fam Physician. 2008;77:339–46.

Pityriasis Versicolor

7. Harada K, Saito M, Sugita T, Tsuboi R. Malassezia species and their associated skin diseases. J Dermatol. 2015;42:250–7.
8. Hu SW, Bigby M. Pityriasis versicolor: a systematic review of interventions. Arch Dermatol. 2010;146:1132–40.
9. Mendez-Tovar LJ. Pathogenesis of dermatophytosis and tinea versicolor. Clin Dermatol. 2010;28:185–9.

Tinea Imbricata

10. Baruzzi RG, Marcopito LF, Vicente LS, Michalany NS. Jorge Lobo's disease (keloidal blastomycosis) and tinea imbricata in Indians from the Xingu National Park, Central Brazil. Trop Dr. 1982;12:13–3.
11. Bonifaz A, Vázquez-González D. Tinea imbricata in the Americas. Curr Opin Infect Dis. 2011;24:106–11.
12. Veraldi S, Giorgi R, Pontini P, Tadini G, Nazzaro G. Tinea imbricata in na Italian child and review of the literatura. Mycopathologia. 2015;180:353–7.

Jorge Lobo's Disease

13. Talhari S, Talhari C. Lobomycosis. Clin Dermatol. 2012;30:420–4.
14. Camargo ZP, Baruzzi RG, Maeda SM, Floriano MC. Antigenic relationship between Loboa loboi and Paracoccidioides braziliensis as shown by serological methods. Med Mycol. 1998;36:413–7.
15. Baruzzi RG, Rodrigues DA, Michalany NS, Salomao R. Squamous-cell carcinoma and lobomycosis (Jorge Lobo's disease). Int J Dermatol. 1989;28:183–5.

Chromoblastomycosis

16. Queiroz-Telles F, de Hoog S, Santos DW, Salgado CG, Vicente VA, Bonifaz A, Roilides E, Xi L, Azevedo CM, da Silva MB, Colombo AL, Walsh TJ. Chromoblastomycosis. Clin Microbiol Rev. 2017;30:233–76.

17. Avelar-Pires C, Simoes-Quaresma JA, Moraes-de Macedo GM, Brasil-Xavier M, Cardoso-de BA. Revisiting the clinical and histopathological aspects of patients with chromoblastomycosis from the Brazilian Amazon region. Arch Med Res. 2013;44:302–6.
18. Jamil A, Lee YY, Thevarajah S. Invasive squamous cell carcinoma arising from chromoblastomycosis. Med Mycol. 2012;50:99–102.

Mycetoma

19. Zijlstra EE, van de Sande WW, Welsh O, Mahgoub el S, Goodfellow M, Fahal AH. Mycetoma: a unique neglected tropical disease. Lancet Inf Dis. 2016;16:100–12.
20. Carrasco-Zuber JE, Navarrette-Dechent C, Bonifaz A, Fich F, Vial-Letelier V, Berroeta-Mauriziano D. Cutaneous involvement in the deep mycoses: a literature review. Part I-subcutaneous mycoses. Actas Dermosifiliogr. 2016;107:806–15.
21. Nenoff P, van de Sande WW, Fahal AH, Reinel D, Schöfer H. Eumycetoma and actinomycetoma – an update on causative agents, epidemiology, pathogenesis, diagnostics and therapy. J Eur Acad Dermatol Venereol. 2015;29:1873–83.

Impetigo

22. Ibrahim F, Khan T, Pujalte GG. Bacterial skin infections. Prim Care. 2015;42:485–99.
23. Oranje AP, de Waard-van der Spek FB. Recent developments in the management of common childhood skin infections. J Infect. 2015;71(Suppl 1):S76–9.
24. Bowen AC, Mahé A, Hay RJ, Andrews RM, Steer AC, Tong SY, Carapetis JR. The global epidemiology of impetigo: a systematic review of the population prevalence of impetigo and pyoderma. PLoS One. 2015;10:e0136789.

Bacterial Folliculitis

25. Lindsei SF, Tosti A. Ethnic hair disorders. Curr Probl Dermatol. 2015;47:139–49.
26. Napierkowski D. Uncovering common bacterial skin infections. Nurse Pract. 2013;38:30–7.
27. Empinotti JC, Uyeda H, Ruaro RT, Galhardo AP, Bonatto DC. Pyodermitis. An Bras Dermatol. 2012;87:277–84.

Erysipelas and Cellulitis

28. Ibrahim F, Khan T, Pujalte GG. Bacterial skin infections. Prim Care. 2015;42:485–99.
29. Oh CC. Cellulitis and erysipelas: prevention. BMJ Clin Evid. 2015.
30. Ralf AB, Kroshinsky D. Cellulitis: a review. JAMA. 2016;316:325–37.

Furunculosis and Abscess

31. Arnáiz-Garcia AM, Arnáiz-Garcia ME, Arnáiz J. Management of furuncle, furunculosis and anthrax. Med Clin (Barc). 2015;144:376–8.
32. Demos M, McLeod MP, Nouri K. Recurrent furunculosis: a review of the literature. Br J Dermatol. 2012;167:725–32.
33. Atanaskova N, Tomecki KJ. Innovative management of recurrent furunculosis. Dermatol Clin. 2010;28:479–87.

Cutaneous Mycobacteriosis

34. Gonzalez-Santiago TM, Drage LA. Nontuberculosis mycobacteria: skin and soft tissue infections. Dermatol Clin. 2015;33:563–77.
35. Zhang X, Liu W, Liu W, Jiang H, Zong W, Zhang G, Jin P, Wang H. Cutaneous infections caused by rapid growing mycobacteria: case reports and review of clinical and laboratory aspects. Acta Dermatol Venereol. 2015;95:985–9.
36. Lamb RC, Dawn G. Cutaneous non-tuberculous mycobacterial infections. Int J Dermatol. 2014;53:1197–204.

Erythrasma

37. Blasco-Morente G, Arias-Santiago S, Pérez-López I, Martinez-López A. Coral-red fluorescence of erythrasma plaque. Sultan Qaboos Univ Med J. 2016;16:e381–2.
38. Pinto M, Hundi GK, Bhat RM, Bala NK, Dandeken S, Martin J, Kambil SM. Clinical and epidemiological features of coryneform skin infections of a tertiary hospital. Indian Dermatol Online J. 2016;7:168–72.
39. Maleki M, Fata A. Evaluation of wood's light and direct smear for diagnosis of pytiriasis versicolor and erithrasma. Saudi Med J. 2005;26:1483–4.

Warts

40. Ockenfels HM. Therapeutic management of cutaneous and genital warts. J Dtsch Dermatol Ges. 2016;14:892–9.
41. Finley C, Korownik C, Kolber MR. What works best for nongenital warts? Can Fam Physician. 2016;62:997.
42. Egawa N, Egawa K, Griffin H, Doorbar J. Human papillomaviruses: epithelial tropisms, and the development of neoplasia. Virus. 2015;7:3363–90.

Focal Epithelial Hyperplasia

43. Kryatova MS, Okoye GA. Dermatology in the north American Indian/Alaska native population. Int J Dermatol. 2016;55:125–34.
44. González-Losa MR, Suarez-Allén RE, Canul-Canche J, Conde-Ferráez L, Eljure-Lopez N. Multifocal epitelial hyperplasiain a community in the Mayan area of Mexico. Int J Dermatol. 2011;50:304–9.
45. Borborema-Santos CM, Castro MM, Santos PJ, Talhari S, Astolfi-Filho S. Oral focal epitelial hyperplasia: report of five cases. Braz Dental J. 2006;17:79–82.

Condyloma Acuminatum

46. Nelson EL, Bogliatto F, Stockdale CK. Vulva intraepithelial neoplasia (VIN) and condylomata. Clin Obstet Gynecol. 2015;58:512–25.

47. Steben M, Garland SM. Genital warts. Best Pract Res Clin Obstet Gynecol. 2014;28:1063–73.
48. Leszczyszyn J, Lebski I, Lysenko L, Hirnle L, Gerber H. Anal warts (condyloma acuminata) – current issues and treatment modalities. Adv Clin Exp Med. 2014;23:307–11.

Molluscum Contagiosum

49. Sterling J. Treatment of warts and molluscum: what does the evidence show? Curr Opin Pediatr. 2016;28:490–9.
50. Afsar FS, Erkan CD, Karaca S. Clinical practice trends in cryosurgery: a retrospective study of cutaneous lesions. Postepy Dermatol Alergol. 2015;32:88–93.
51. Shisler JL. Immune evasion strategies of molluscum contagiosum virus. Adv Virus Res. 2015;92:201–52.
52. McCollum AM, Holman RC, Hughes CM, Mehal JM, Folkema AM, Redd JT, Cheek JE, Damon IK, Reynolds MG. Molluscum contagiosum in a pediatric American Indian population: incidence and risk factors. PLoS One. 2014;9:e103419.

Herpes Simplex

53. Sauerbrei A. Herpes genitalis: diagnosis, treatment and prevention. Geburtshilfe Frauenheilkd. 2016;76:1310–7.
54. Grinde B. Herpesviruses: laency and reactivation – viral strategies and host response. J Oral Microbiol. 2013.
55. Kumarasamy N, Balakrishnan P,Venkatesh KK, Srikrishnan AK, Cecelia AJ, Thamburaj E, Solomon S, Mayer KH. Prevalence and incidence of sexually transmitted infections among South Indians at increased risk of HIV infection. AIDS Patient Care STDs. 2008;22:677–82.

Herpes Zoster

56. Mallick-Searle T, Snodgrass B, Brant JM. Postherpetic neuralgia: epidemiology, pathophysiology, and pain management pharmacology. J Muldiscip Healthc. 2016;9:447–54.
57. Peterson N, Goodman S, Peterson M, Peterson W. Herpes zoster in children. Cutis. 2016;98:94–5.
58. Schmader K. Herpes zoster. Clin Geriatr Med. 2016;32:539–53.

Scabies

59. Romani S, Steer AC, Whitfeld MJ, Kaldor JM. Prevalence of scabies and impetigo worldwide: a systematic review. Lancet Infect Dis. 2015;15:960–7.
60. Dadabhoy I, Butts JF. Parasitic skin infections for primary care physicians. Prim Care. 2015;42:661–75.
61. Fuller LC. Epidemiology of scabies. Curr Opin Infect Dis. 2014;26:123–6.

Pediculosis

62. Bohl B, Evetts J, McClaim K, Rosenauer A, Stellitano E. Clinial practice update: pediculosis capitis. Pediatr Nurs. 2015;42:277–34.
63. Muhammad Zayyid M, Saidatul Saadah R, Adil AR, Rohela M, Jamaiah I. Prevalenece of scabies and head lice among children in a welfare home in Pulau Pilang, Malaysia. Trop Biomed. 2010;27:442–6.
64. Boutellis A, Abi-Rached L, Raoult D. The origin and distribution of human lice in the world. Infec Genet Evol. 2014;23:209–17.

Cutaneous Larva Migrans

65. Prickett KA, Ferringer TC. What's eating you?. Cutaneous larva migrans. Cutis. 2015;95:126–8.
66. Eichelmann K, Tomecki KJ, Martínez JD. Tropical dermatology: cutaneous larva migrans, gnathostomiasis, cutaneous amebiasis and trombiculiasis. Semin Cutan Med Surg. 2014;33:133–5.
67. Meotti CD, Plates G, Nogueira LL, Silva RA, Paolini KS, Nunes EM, Bernardes Filho F. Cutaneous larva migrans on the scalp: atypical presentation of a common disease. An Bras Dermatol. 2014;89:332–3.

Tungiasis

68. Feldmeier H, Heulkelbach J, Ugbomoiko US, Sentongo E, Mbabazi P, von Samson-Himmelstjerna G, Krantz I, International expert group for Tungiasis. Tungiasis-a neglected disease with many challenges for global public health. PLoS Negl Trop Dis. 2014;8:e3133.
69. Vallarelli AF, Souza EM. Disseminated tungiasis. An Bras Dermatol. 2011;86:1027–8.
70. Miller H, Rodríguez G. Tungiasis in native Amerindians in Vaupés province; epidemiology, clinical aspects, treatment, and prevention. Biomedica. 2010;30:215–37.

Tegumentary Leishmaniasis

71. Oryan A, Akbari M. Worldwide risk factors in leishmaniasis. Asian Pac J Trop Med. 2016;9:925–32.
72. Kaye PM, Beattie L. Lesson from other diseases: granulomatous inflammation in leishmaniasis. Semin Immunopathol. 2016;38:249–60.
73. Costa-Silva MF, Gomes LI, Martins-Filho OA, Rodrigues-Silva R, Freire J de M, Quaresma PF, Pascoal-Xavier MA, Mendes TA, Serakides R, Zauli DA, Campi-Azevedo CA, Melo MN, Gontijo CM, Peruhype-Magalhães V, Teixeira-Carvalho A. Gene expression profile of cytokines and chemokines in skin lesions from Brazilian Indians with localized leishmaniasis. Mol Immunol. 2014;57:74–85.

5 Inflammatory Diseases

5.1 Pityriasis Alba and Eczematid

Pityriasis alba and eczematid are recurrent, noninfectious, and nontransmissible skin diseases of a constitutional nature. They are frequent among atopic (allergic) patients with dry skin. There is spontaneous remission, but recurrence is common. Eczematid may be considered an inflammatory form of pityriasis alba. It is a common disease, more frequent among children and young adults. Allergic people (with rhinitis or allergic bronchitis) are more likely to have the disease. Pityriasis alba is more clinically evident following sun exposure without sunscreen, which increases the contrast of the lesion against normal skin. Pityriasis alba shows white spots with fine scaling. This form affects the face (malar region), arms, and trunk (Figs. 5.1, 5.2, and 5.3). It is asymptomatic. Eczematid is characterized by rounded erythematous-scaly lesions that can affect any part of the body but are more common in the trunk and limbs (Figs. 5.4 and 5.5). This form is asymptomatic or can cause slight itching.

Diagnosis is clinical. In pityriasis alba, the histopathology shows a reduction in melanin and melanocytes in the basal layer of the epidermis. In eczematid, direct mycological examination is recommended if there is doubt about colonization by fungus.

The recommendations are to avoid prolonged exposure to the sun, take long baths, and to use soap extensively. The use of moisturizers is very useful for skin dryness control. In the case of pruritic lesions, the use of low-potency corticosteroids, such as 1% hydrocortisone cream, is recommended.

Frequently, indigenous populations are exposed to the sun, provoking a high contrast of pityriasis alba lesions and adjacent normal skin.

5.2 Contact Dermatitis

Contact dermatitis is an eczematous disease caused by contact with some substance that causes irritation or allergy in a given area. The patient may or may not have previous exposure to the substance (allergen).

The increase in chemicals has led to increased chances of developing contact dermatitis. This disease can affect anyone at any age, with or without prior exposure to the causative substance.

The lesions that arise in association with this disease are very itchy. When they are acute, vesicles or blisters are observed, accompanied by erythema and edema. Because they itch a lot, they are easily traumatized by scratching and become secretory. Chronic lesions are dry, and they are scaly with thicker and itchy skin. Lesions occur in areas where there was contact with the suspected substance (e.g., feet, hands, face) (Figs. 5.6, 5.7, and 5.8). In allergic contact dermatitis, the lesions may reach other regions and become disseminated.

Once the suspect substance is identified, the patient should avoid any contact with or handling of the allergen. In acute cases, the use of topical corticosteroid (dexamethasone, betamethasone) until clinical improvement is recommended. The use of systemic antihistamines (dexchlorpheniramine, hydroxyzine, loratadine) can help to control the pruritus. In more severe cases, the use of systemic corticosteroids (injectable betamethasone, oral prednisone) is useful. In chronic cases, better results are obtained with a topical corticosteroid (betamethasone, clobetasol) as an ointment associated with oral antihistamine.

Indigenous populations have access to unknown and different substances. They can lead to various conditions of contact dermatitis as a primary irritant or allergic substance.

M.C. Florian et al., *Dermatological Atlas of Indigenous People*, DOI 10.1007/978-3-319-59446-0_5

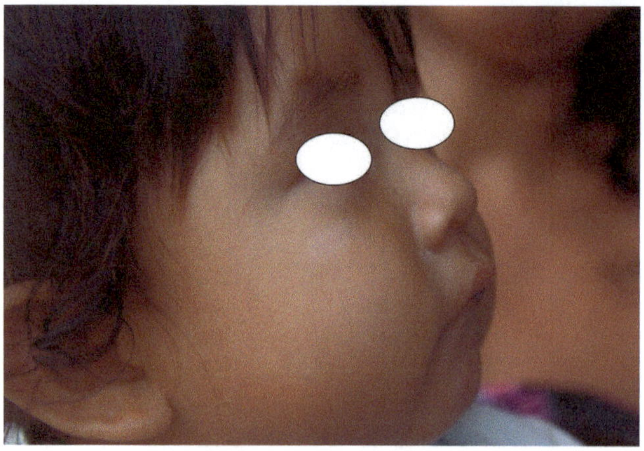

Fig. 5.1 Pityriasis alba: lesions on face

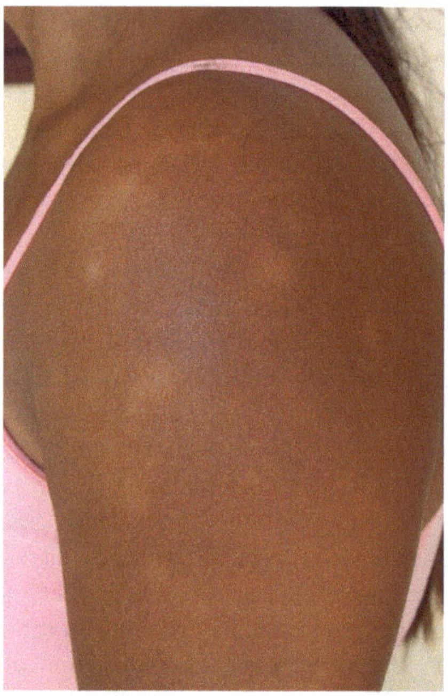

Fig. 5.2 Pityriasis alba: lesions on arm

5.3 Seborrheic Dermatitis

Seborrheic dermatitis is a noninfectious and nontransmissible chronic skin disease. It is often associated with allergies (asthma, rhinitis, sinusitis, or eczema). On the scalp, the disease develops scales known as dandruff. There are many related factors such as fungal colonization (*Malassezia* spp.), sebaceous secretions, and emotional stress.

It is a disease with worldwide distribution and is frequent during the first 6 months of life and in adults after adolescence.

There are reddish (erythematous) lesions, with greasy scales, that can affect the scalp, face (eyebrow and nose region), chest, groin, and genital region (Figs. 5.9, 5.10, 5.11,

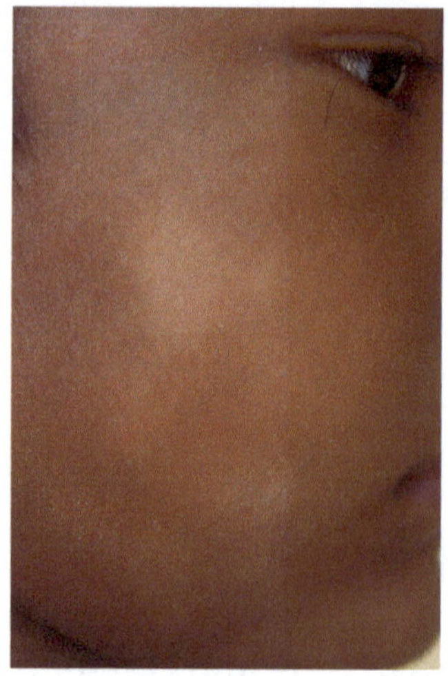

Fig. 5.3 Pityriasis alba: lesions on face

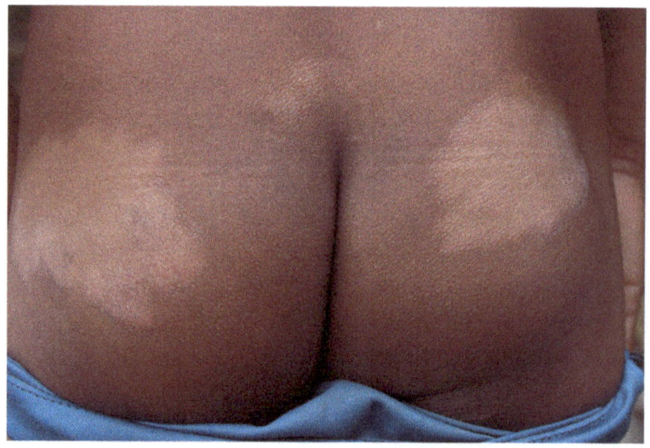

Fig. 5.4 Eczematid: lesions on buttocks

and 5.12). These lesions can cause local itching and inflammation. In young children (under 6 months of age) they can reach other areas of the skin as well (face, neck, trunk, and extremities). Often the formation of scales on the vertex of the scalp are observed, depending on the extent of the skin lesions. Mild forms are treated with topical azoles or ciclopirox olamine. Topical 1% hydrocortisone cream twice daily can be used until the inflammatory symptoms are under control. In adults with extensive lesions, betamethasone cream 0.1% twice daily is useful. The combination of ketoconazole, zinc pyrithione, selenium sulfide, or ciclopirox olamine shampoo may aid in treatment. The crusty mass on a baby's scalp can be removed with sweet almond oil.

5.4 Neurodermatitis

Neurodermatitis is also called *lichen simplex chronicus*. It is a thickening of the skin resulting from chronic itching by different external stimuli, especially eczema or dermatitis (e.g., chronic contact dermatitis) and insect bites.

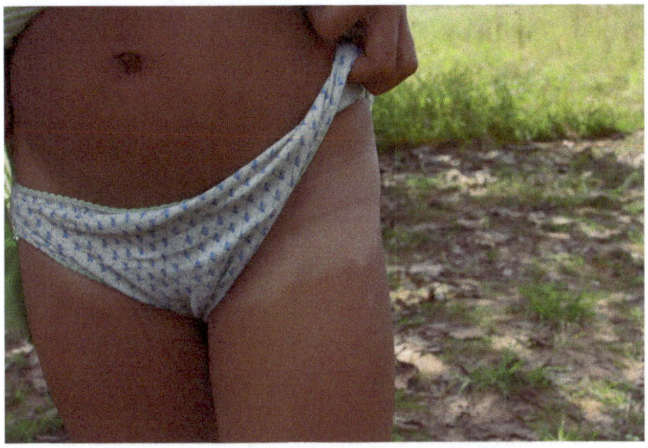

Fig. 5.5 Eczematid: lesions on lower limbs

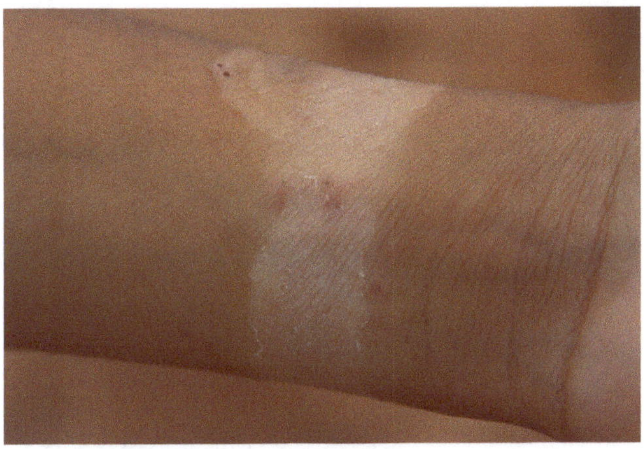

Fig. 5.6 Contact dermatitis: lesions on wrist

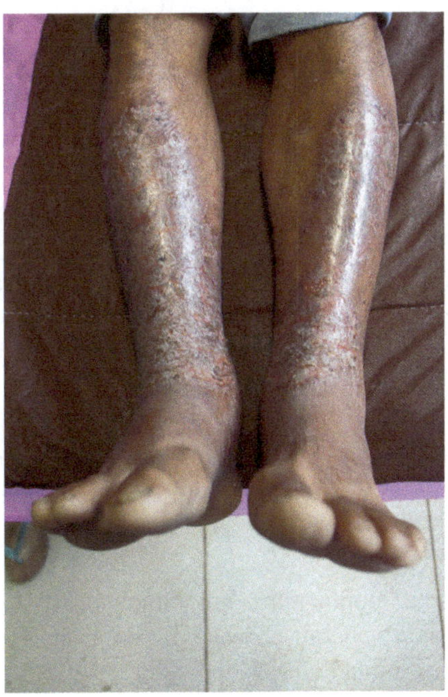

Fig. 5.7 Contact dermatitis: lesions on legs

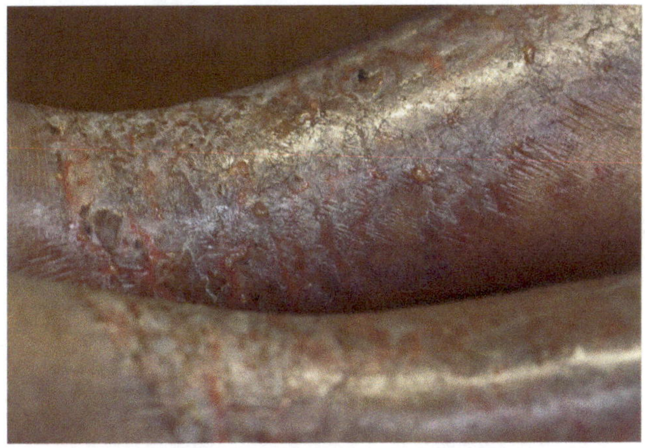

Fig. 5.8 Contact dermatitis: lesions on legs

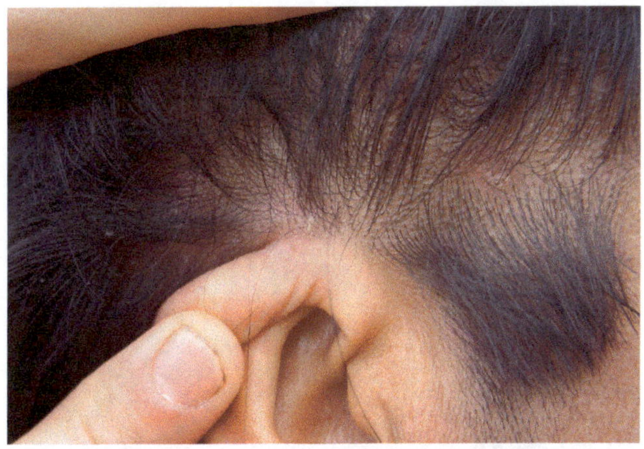

Fig. 5.9 Seborrheic dermatitis: lesions on scalp

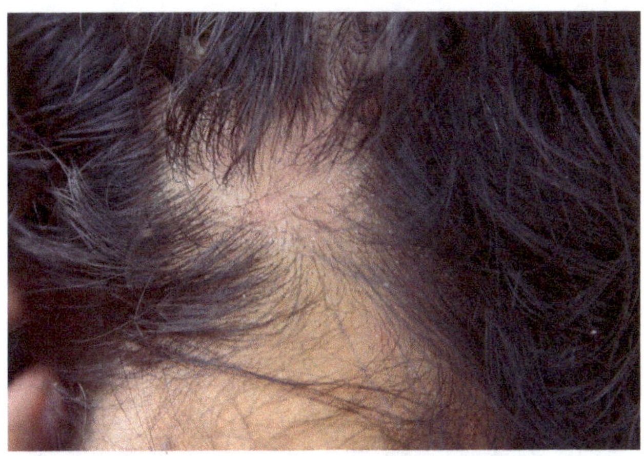

Fig. 5.10 Seborrheic dermatitis: lesions on scalp

It afflicts people exposed to environmental conditions that lead to contact dermatitis, insect bites, or other chronic dermatoses. It affects anyone, especially elderly patients with dry skin. Emotional stress is another factor that can induce a pruritic sensation and scratching (Figs. 5.13, 5.14, and 5.15).

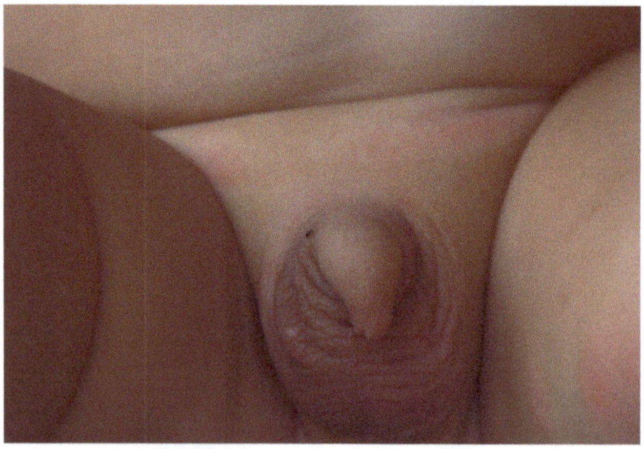

Fig. 5.11 Seborrheic dermatitis: lesions on genital region

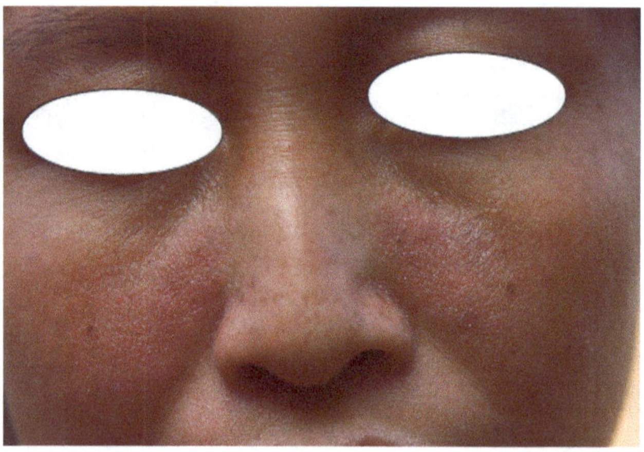

Fig. 5.12 Seborrheic dermatitis: lesions on face

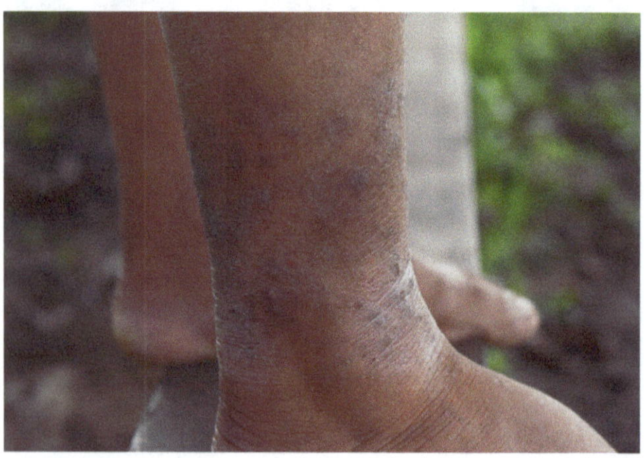

Fig. 5.13 Neurodermatitis: lesions on leg

Thick, scaly, and hyperchromic plaques arise, showing accentuation of natural lines on the skin surface (lichenification). Pruritus is always present and may be mild, moderate,

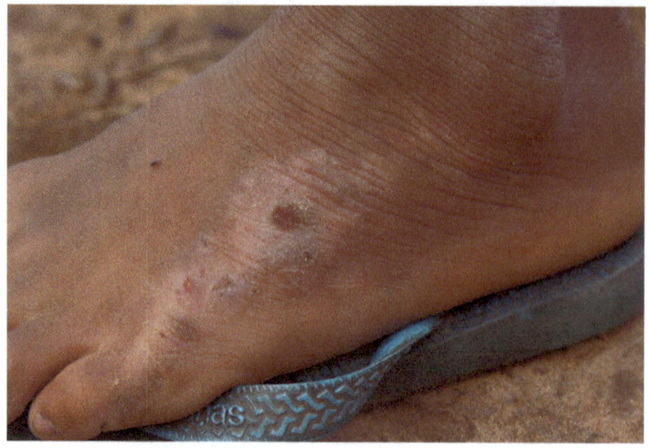

Fig. 5.14 Neurodermatitis: lesions on foot

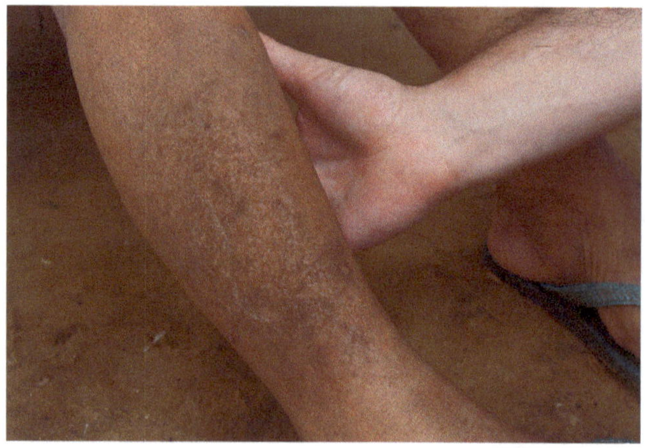

Fig. 5.15 Neurodermatitis: lesions on leg

or very intense. Solitary or multiple lesions are observed on the posterior region of the neck and dorsum, in the anogenital region, and on extremities. Diagnosis is clinical. A skin biopsy may be necessary when there is diagnostic doubt. Examinations of the removed skin fragment are performed based on clinical suspicions (e.g., histopathological examination, cultures).

The pruritus (itch) should be blocked with topical corticosteroids, preferably in ointment with or without occlusion. Intralesional infiltration with triamcinolone may be performed. The combination with systemic antihistamines is useful.

After improvement, remove the possible triggering cause (e.g., contact dermatitis) and start moisturizing the skin with creams or oils; otherwise the condition may flare up frequently.

Indigenous populations are continuously exposed to insect bites, which can provoke a chronic state of itching. After several months, neurodermatitis plaques can develop and affect the quality of life.

5.5 Psoriasis

Psoriasis is a dermatosis characterized by erythematous and desquamative lesions. Its evolution is chronic, and its etiology is not well understood, but it is believed to involve genetic, biochemical, immune, and environmental factors. There is an increase of Th1 (T cell helper) activity, including Th17. Nowadays, it is known that psoriasis increases the risk for metabolic syndrome and cardiovascular disease. It can affect any age group, both men and women alike. In approximately 30% of occurrences, we find familial cases of the disease. Psoriasis is associated with HLA-Cw6, and it is related to early onset of the disease.

There are several forms of clinical presentation. The classic lesions are well-defined erythematous-desquamative plaques of different sizes. The distribution is symmetrical, affecting mainly the scalp, elbows, knees, and sacral region (Figs. 5.16, 5.17, 5.18, and 5.19). Lesions can appear anywhere on the body, including on the nails (Fig. 5.20), in oral mucosa, and in joints (psoriatic arthritis). Psoriasis is also related to metabolic syndrome and atherosclerotic cardiovascular diseases. Diagnosis is made by clinical presentation. When necessary, a skin biopsy is performed for histopathological examination.

Topical corticosteroids and sun exposure (ultraviolet radiation or phototherapy) are used for mild forms. There are other topical therapeutic options (coaltar, calcipotriol) and systemic medications (methotrexate, cyclosporine A, acitretin). Recently a large number of biologic therapies have been developed, such as infliximab, etanercept, adalimumab, ustekinumab, and others. Routine laboratory controls should be performed.

There is one publication analyzing the Australian indigenous population. Heyes et al. showed that psoriasis is uncommon in this population and occurs as skin cancer. Analysis of an indigenous tribe from Taiwan revealed a high frequency of DRB1 alleles that is also found in people from Oceania, Australian aborigines, and American Indians. The allele DRB1*1401 is frequent in most tribes but is not related to psoriasis in these populations, as previously shown in some publications.

5.6 Insect Bites

Reactions to insect bites are a type of prurigo (papules with a small vesicle and itching). These reactions occur worldwide, although more frequently in tropical areas, near rivers and forests. The first reaction related to insect bites is mainly mediated by histamine and kinins. This reaction disappears quickly. After an immunologic sensitization to these bites, there is a more extensive skin manifestation that could be

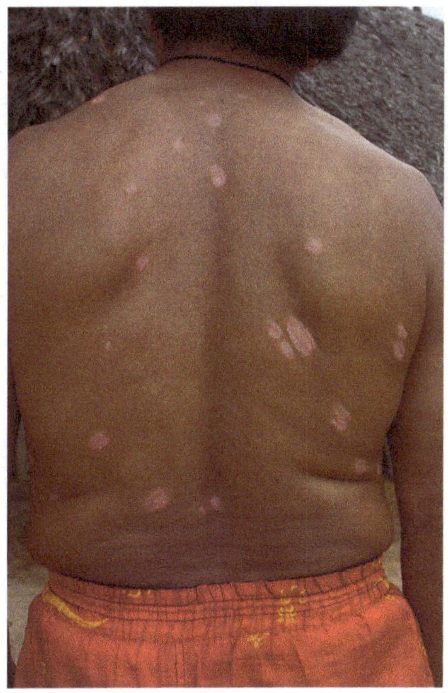

Fig. 5.17 Psoriasis: lesions on dorsum

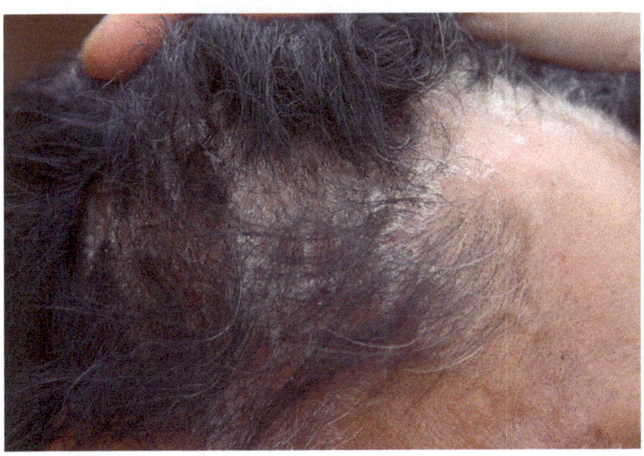

Fig. 5.16 Psoriasis: lesions on scalp

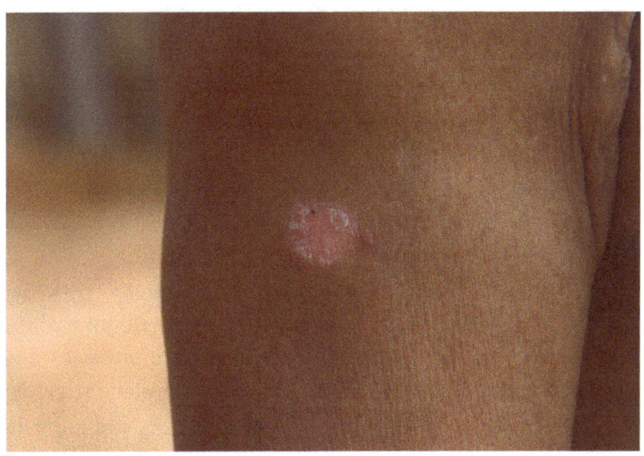

Fig. 5.18 Psoriasis: typical lesion with erythema and scales

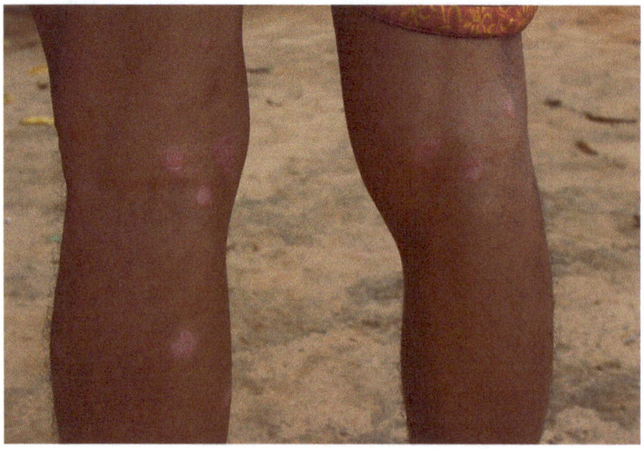

Fig. 5.19 Psoriasis: lesions on legs

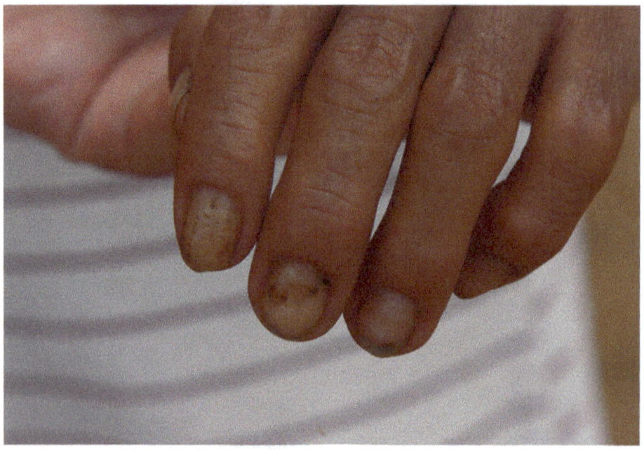

Fig. 5.20 Psoriasis: pitting lesions on nails

infected. Insect bite reactions are more common in children, especially among atopic patients (allergic rhinitis, asthma, or bronchitis).

Small erythematous papules and edematous pruritic lesions are distributed on the head and extremities (Figs. 5.21 and 5.22). They usually have a central crust and may lead to secondary bacterial infection. Grouped or disseminated lesions develop following localized insect bites (Figs. 5.23 and 5.24). Chronic hyperpigmented papules and nodules appear, resembling prurigo nodularis.

Diagnosis is clinical. Skin biopsy for pathological examination could be performed if the diagnosis is uncertain. In mild cases, topical corticosteroids (cream, ointment, lotion) could be prescribed. In widespread manifestations, systemic antihistamines would be necessary, with or without a short course of systemic corticosteroids. Insect bite prevention is very important, including through the use of protective clothing and repellents. Exposure to insects is very common in tropical areas as a consequence of siting traditional houses near water and of poverty.

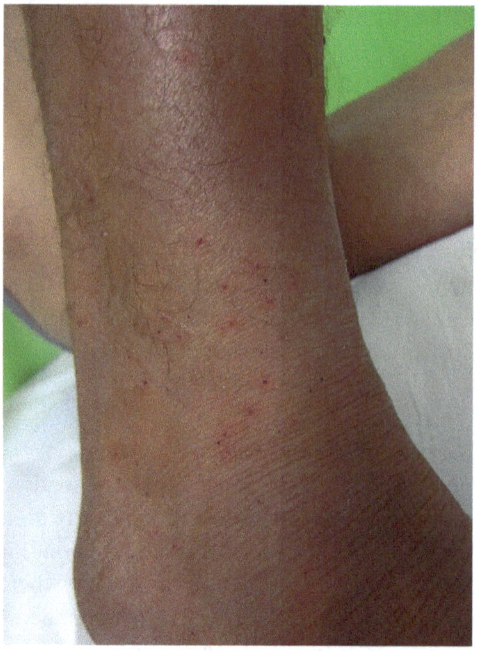

Fig. 5.21 Insect bites: lesions on leg

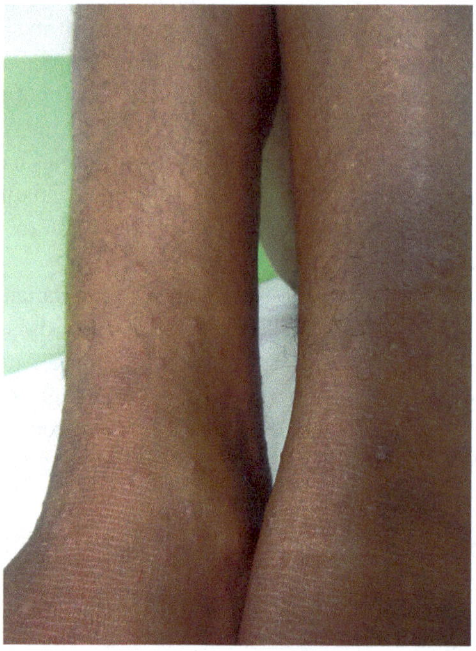

Fig. 5.22 Insect bites: lesions on leg

5.7 Urticaria

Urticaria is the sudden appearance of swelling and erythematous lesions. The lesions are very pruritic and disappear within 24 h. They can vary in size, from a few millimeters to large plaques. Possible triggers causing this cutaneous reaction are numerous (e.g., drugs, foods, infections, internal diseases).

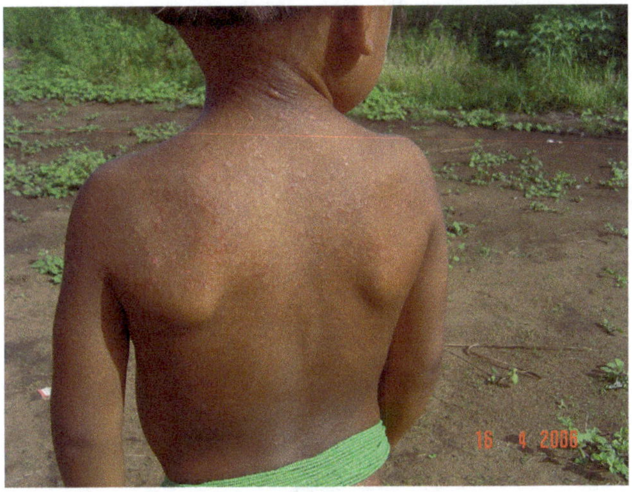

Fig. 5.23 Insect bites: disseminated lesions

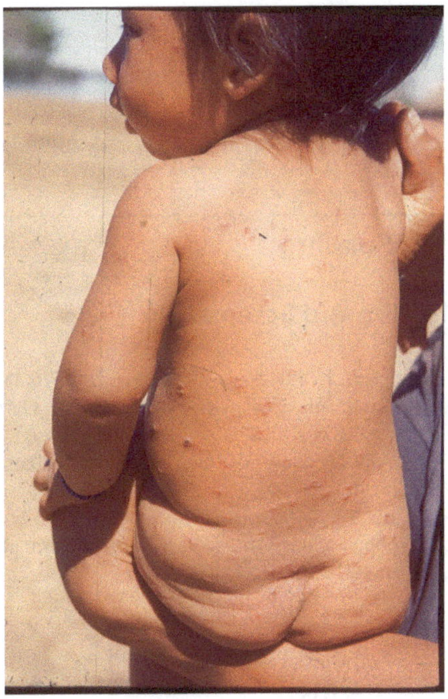

Fig. 5.24 Insect bites: disseminated lesions

Angioedema occurs in the subcutaneous region and in the deep dermis and involves especially the face. Urticaria affects people of all ages and is more frequent in females. There are family cases.

Erythematous, edematous lesions of various sizes (from a few millimeters to giant lesions that can affect any body part) (Figs. 5.25 and 5.26). Lesions tend to suddenly appear and disappear in a few minutes or, in some cases, a few hours and leave no scar. The pruritus (itch) is very intense. Edema can reach deeper layers of the skin, especially the eyelids, lips, tongue, and larynx (angioedema). It may be accompanied by difficulty breathing and constitute a medical emergency. Urticaria is classified as acute and chronic.

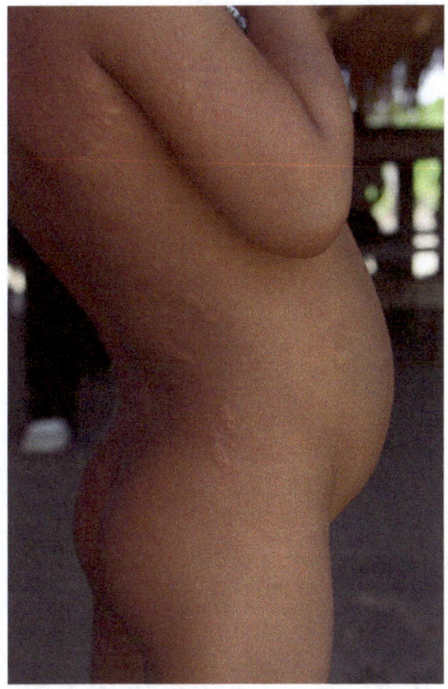

Fig. 5.25 Urticaria: disseminated edematous lesions

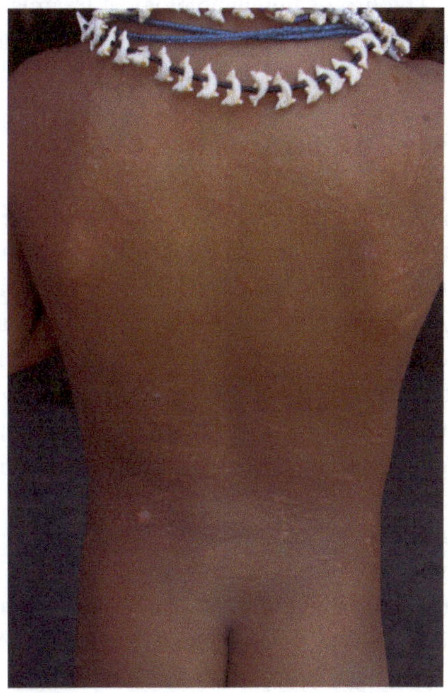

Fig. 5.26 Urticaria: disseminated edematous lesions

The latter presents after a period of 6 weeks or more. Common urticaria appears spontaneously, and the cause is not defined. Physical urticaria is very rare and is induced by mechanical stimuli, temperature variations, exercise (cholinergic), and water (aquagenic).

Diagnosis is made based on the clinical presentation. It is rarely necessary to perform a skin biopsy for pathological examination. Initially look for the cause and remove it.

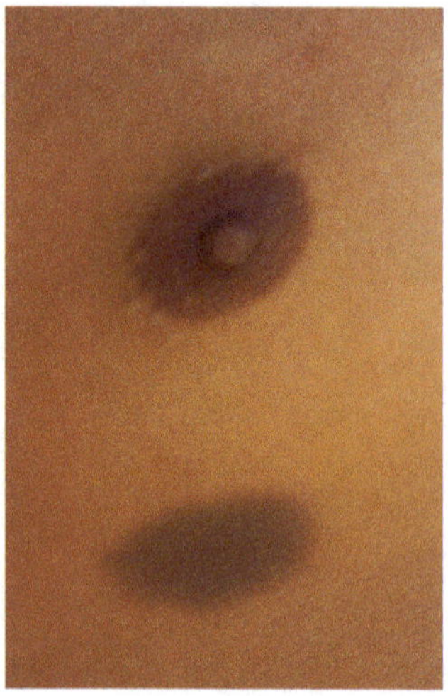

Fig. 5.27 Fixed drug eruption: oval hyperchromic lesion

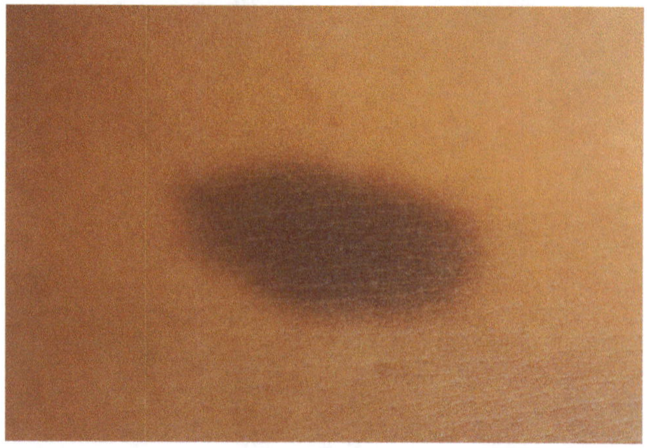

Fig. 5.28 Fixed drug eruption: oval hyperchromic lesion

In milder cases, use systemic antihistamines (hidroxizine, loratadine, H_2 antagonists). In more severe cases, oral corticosteroids can be associated. In emergencies (difficulty breathing), use subcutaneous epinephrine and tracheotomy as needed. The use of immunobiological treatment is difficult because indigenous people live far from hospitals.

5.8 Fixed Drug Eruption

Drugs can cause adverse reactions in the skin. One is a fixed drug eruption that develops 1 or 2 weeks after the first exposure to the drug. Continuous intake and re-exposure to the same drug is responsible for skin lesion multiplication. The disease can start at any age and affect different parts of the body. Different drugs can show the same type of injury. The most common offending drugs are sulfonamides, nonsteroid anti-inflammatory drugs, barbiturates, tetracyclines, and carbamazepine.

One or more erythematous patches are observed that become hyperchromic and that arise when there is exposure to the drug responsible for the reaction. In general, the lesions are round or oval (Figs. 5.27 and 5.28). With new exposure, the lesions reappear in the same locations, adding new lesions. They can appear anywhere on the skin but have a certain preference for the palms, soles of the feet, and mucous membranes. The lesions tend to fade with time, but in some cases, they become permanent. Pigmentation is due to dermal melanophage infiltration.

Diagnosis is clinical. Biopsy for pathological examination should be done when there is diagnostic uncertainty. Identify the offending drug and cease using it. If the lesions are erythematous or bullous, treatment with topical or systemic corticosteroids is recommended. Nowadays indigenous populations have access to conventional drugs, and cutaneous adverse drug reactions should be included in differential diagnoses. Consideration should also be given to the possibility of reactions to any medicinal herbs that might be used.

Suggested Reading

Pityriasis Alba and Eczematid

1. Lin RL, Janniger CK. Pityriasis alba. Cutis. 2005;76(1):21–4.
2. Miazek N, Michalek I, Pawlowska-Kisiel M, Olszewska M, Rudnicka L. Pityriasis alba – common disease, enigmatic entity: up-to-date review of the literature. Pediatr Dermatol. 2015;32(6):786–91.
3. Patel AB, Kubba R, Kubba A. Clinicopathological correlation of acquired hypopigmentary disorders. Indian J Dermatol Venereol Leprol. 2013;79(3):376–82.

Contact Dermatitis

4. Gupta D, Thappa DM. Dermatoses due to Indian cultural practices. Indian J Dermatol. 2015;60(1):3–12.
5. Mortimer S, Reeder M. Botanicals in dermatology: essential oils, botanical allergens, and current regulatory practices. Dermatitis. 2016;27(6):317–24.
6. Nyguyen V, Simon L, Jaqua E. Allergic dermatoses. Prim Care. 2016;43(3):433–49.

Seborrheic Dermatitis

7. Borda LJ, Wikramanayake TC. Seborrheic dermatitis and dandruff: a comprehensive review. J Clin Investig Dermatol. 2015;3(2). doi:10.13188/2373-1044.1000019. Epub 2015 Dec 15. https://www.ncbi.nlm.nih.gov/pmc/articles/PMC4852869/.

8. Clark GW, Pope SM, Jaboori KA. Diagnosis and treatment of seborrheic dermatitis. Am Fam Physician. 2015;91(3):185–90.
9. Naldi L, Diphoorn J. Seborrhoeic dermatitis of the scalp. BMJ Clin Evid. 2015;2015:1713.

Neurodermatitis

10. Kanto R, Thyssen JP, Paller AS, Silverberg JI. Atopic dermatitis, atopic eczema, or eczema? a systematic review, meta-analysis, and recommendation for uniform use of 'atopic dermatitis'. Allergy. 2016;71(10):1480–5.
11. Lotti T, Buggiani G, Prignano F. Prurigo nodularis and lichen simplex chronicus. Dermatol Ther. 2008;21(1):42–6.
12. Weisshaar E, Fleischer AB Jr, Bernhard JD, Cropley TG. Pruritus and dysesthesia. In: Bolognia JL, Jorizzo JL, Schaffer JV, editors. Dermatology. 3rd ed. Amsterdam: Elsevier Sauders; 2012. p. 111–25.

Psoriasis

13. Chu CC, Lin M, Nakajima F, et al. Diversity of HLA among Taiwan's indigenous tribes and the Ivatans in the Philippines. Tissue Antigens. 2001;58(1):9–18.
14. Heyes C, Tait C, Toholka R, Gebauer K. Non-infectious skin disease in indigenous Australians. Australas J Dermatol. 2014;55(3):176–84.
15. Singh S, Young P, Armstrong AW. Relationship between psoriasis and metabolic syndrome: a systematic review. G Ital Dermatol Venereol. 2016;151(6):663–77.

Insect Bites

16. Lee H, Halverson S, Mackey R. Insect allergy. Prim Care. 2016;43(3):417–31.
17. Naafs B. Allergic skin reactions in the tropics. Clin Dermatol. 2006;24(3):158–67.
18. Sicherer SH, Leung DY. Advances in allergic skin disease, anaphylaxis, and hypersensitivity reactions to foods, drugs, and insects in 2014. J Allergy Clin Immunol. 2015;135(2):357–67.

Urticaria

19. Amaya D, Sánchez A, Sánchez J. Inducible urticaria: case series and literature review. Biomedica. 2016;36(1):10–21.
20. Naafs B. Allergic skin reactions in the tropics. Clin Dermatol. 2006;24(3):158–67.
21. Nguyen V, Simon L, Jaqua E. Allergic dermatoses. Prim Care. 2016;43(3):433–49.

Fixed Drug Eruption

22. Flowers H, Brodell R, Brents M, Wyatt JP. Fixed drug eruptions: presentation, diagnosis, and management. South Med J. 2014;107(11):724–7.
23. Patel TK, Thakkar SH, Sharma D. Cutaneous adverse drug reactions in Indian population: a systematic review. Indian Dermatol Online J. 2014;5(suppl 2):S76–86.
24. Shiohara T. Fixed drug eruption: pathogenesis and diagnostic tests. Curr Opin Allergy Clin Immunol. 2009;9(4):316–21.

Cysts, Benign and Malignant Neoplastic Diseases

6

6.1 Seborrheic Keratosis

Seborrheic keratosis is a common benign proliferation of epidermal cells. This dermatosis can occur in any skin phototype; however, the pigmented form is more frequent in high skin phototypes.

There is no predilection for sex, and older populations (over 40 years) are affected. Lesions develop with progression in size and number. Sun exposure and age are implicated factors.

The clinical presentation consists of keratotic and waxy papules and plaques with horn pseudocysts. A papillomatous or verrucous surface is very common, and its colors range from rose to brown-black (Fig. 6.1). The papules and plaques mainly affect the face and trunk, although isolated lesions may develop on extremities. There is a flat macular form that occurs mainly on the face, which can be mistaken for solar melanosis.

Diagnosis is clinical; however, histopathology is quite characteristic with hyperkeratosis and proliferation of squamous and basaloid cells. The presence of horn pseudocysts is characteristic. Lesions are treated by chemical cauterization (trichloroacetic acid), electrodessication, and cryotherapy (liquid nitrogen). Surgical shave excision is another possibility. Like the Asian population, the indigenous population presents the pigmented macular form. During their evolution, these lesions become verrucous or papillomatous.

6.2 Melanocytic Nevi

Melanocytic nevi are derived from modified melanocytes (cells responsible for skin color). They can be congenital (present from birth) or acquired (arise over the course of one's life). They are related to genetic and environmental factors (exposure to sunlight, in the case of acquired nevi) and are commonly called *signs* or *speckles*.

The epidemiological importance of nevi is associated with some risk of transformation into a type of skin cancer called cutaneous melanoma, but the assessment of this risk takes into account many factors that should be analyzed for each situation (such as skin type, age, family history of melanoma). Melanocitic nevi are well circumscribed, brownish or blackish colored, and round or ovoid in shape. They may range in size from a few millimeters to several centimeters (Figs. 6.2, 6.3, 6.4, and 6.5).

Clinical signs requiring the attention of health professionals are asymmetry, irregular edges, many different colors in the same lesion, a size greater than 6 mm, and rapid growth. These indicate the initial changes into atypical nevi or melanoma. When there is a suspicion that a nevus is atypical, a biopsy of the patient's lesion should be taken for pathological examination. In most cases no treatment is necessary, only clinical follow-up. When there is suspicion of transformation to melanoma, excisional biopsy is indicated. Among indigenous populations, special attention is required for acral lentiginous melanoma and atypical nevi on the soles and palms. In spite of intense exposure to the sun without sunscreen protection, there have been no reports of melanoma in Xingu Indigenous Park, showing that genetic factors are very important for melanoma onset.

6.3 Nevus Depigmentosus

Nevus depigmentosus is a congenital cutaneous condition characterized by irregularly shaped hypopigmented spots of varying sizes and well-defined borders (Figs. 6.6). They can be located at or follow the lines of Blaschko. The condition may be less evident at birth and better visualized later. Most of the time it appears in the first 3 years of life. Examination under Wood's light will reveal the spots.

Nevus depigmentosus is usually an isolated condition of the skin but may be related to other changes (neurological, orthopedic). Differential diagnosis should be made with vitiligo and nevus anemicus. In countries where leprosy is

M.C. Florian et al., *Dermatological Atlas of Indigenous People*, DOI 10.1007/978-3-319-59446-0_6

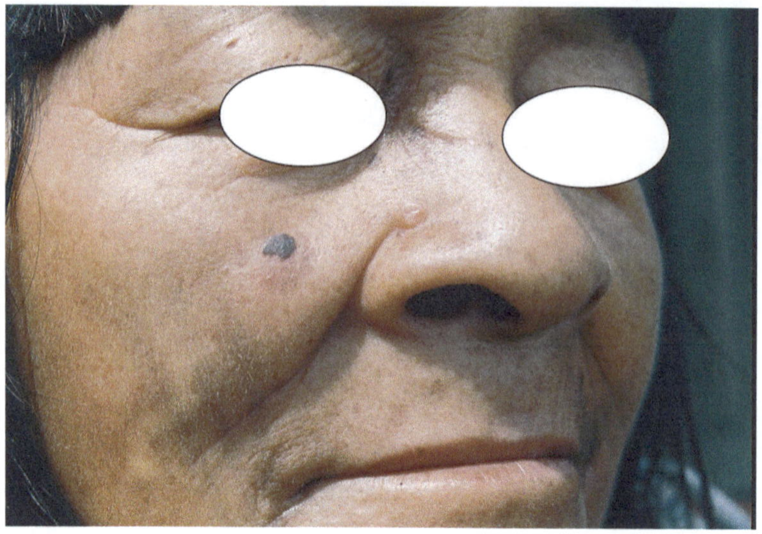

Fig. 6.1 Seborrheic keratosis: single keratotic lesion on face

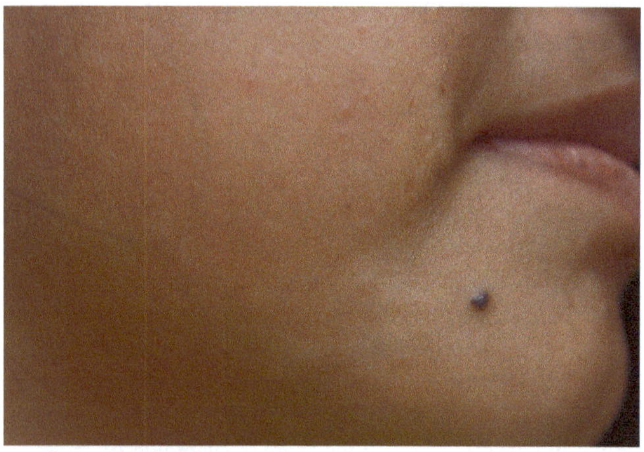

Fig. 6.2 Melanocytic nevi: lesions on face

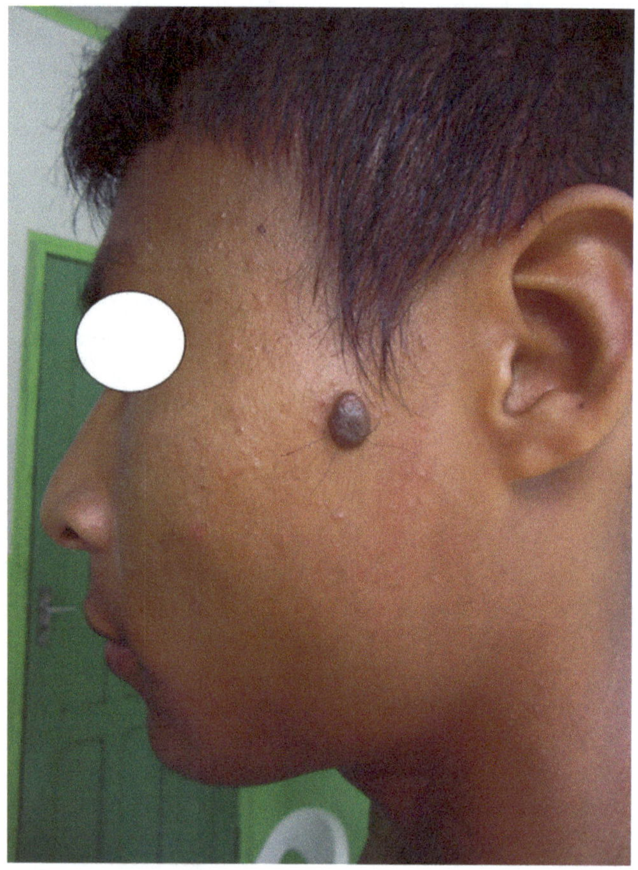

Fig. 6.3 Melanocytic nevi: lesion on face

endemic, a differential diagnosis should be made with indeterminate leprosy, and a thermal, pain, and tactile sensibility test should be performed. A skin biopsy for anatomopathological examination may be performed if necessary. There will be a decrease in melanin, but there is usually no decrease in the number of melanocytes.

Therapeutic options (laser therapy or melanocyte transplantation) can be indicated if aesthetic impairment is very significant.

6.4 Mucous Cyst

A mucous cyst of the oral mucosa is also called a mucocele. It occurs through damage to minor salivary ducts, resulting in mucoid secretion accumulation in the oral mucosa. Mucoceles are seen more frequently on lower labial mucosa, but they also appear on the floor of the mouth and tongue. There is no preference for age or sex.

A mucous cyst is a dome-shaped, soft, and translucent papule or nodule, usually located on the lower lip mucosa (Fig. 6.7). They can range in size from several millimeters to

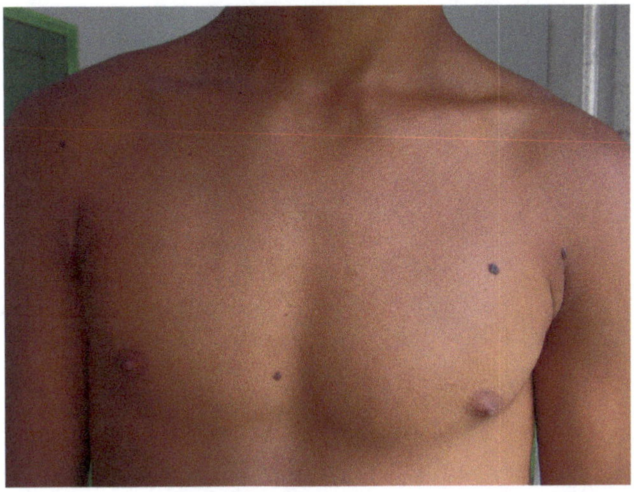

Fig. 6.4 Melanocytic nevi: lesions on chest

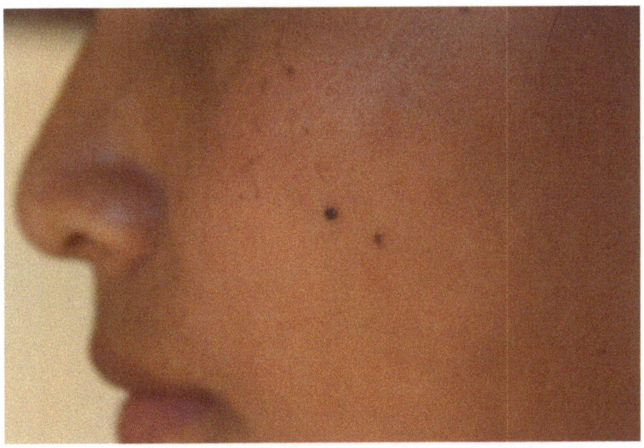

Fig. 6.5 Melanocytic nevi – lesions on the face

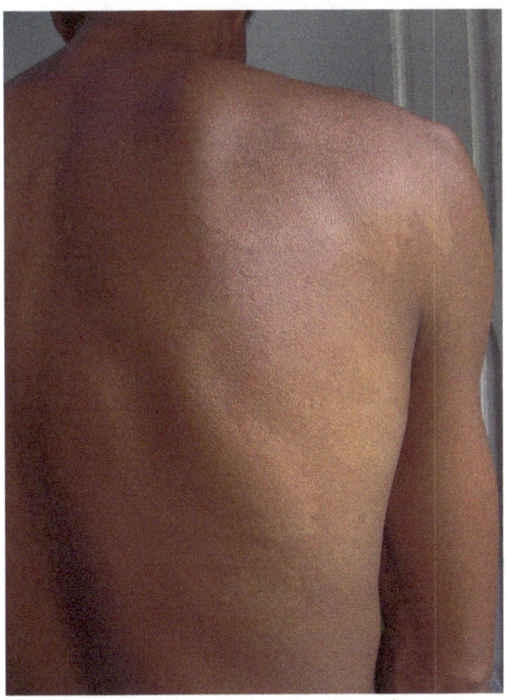

Fig. 6.6 Nevus depigmentosus: lesions on dorsum

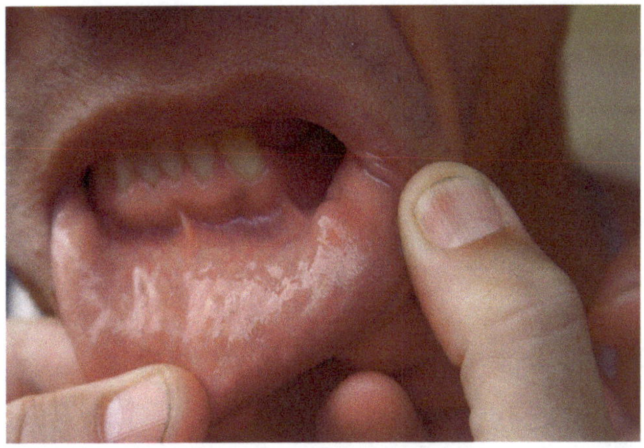

Fig. 6.7 Mucous cyst: lesion on lower lip

over 1 cm. When the cyst disrupts, a transparent and viscous liquid (sialomucin) is eliminated.

A biopsy for pathological examination should be done when the clinical diagnosis is uncertain. Lesions may be removed surgically. Cryotherapy and infiltration with corticosteroids are therapeutic alternatives.

6.5 Pigmented Basal Cell Carcinoma

Basal cell carcinoma is the most common skin cancer in the general population. There are several clinical presentations, one of which is pigmented basal cell carcinoma. Chronic sun exposure is the main cause of the development of the cancer, but genetic factors are also important. This skin cancer has local growth, with a reduced chance of spreading (metastases). It is curable with appropriate treatment.

The disease affects more fair-skinned people, especially those older than 40 who have a history of many years of sun exposure or family history of the disease. In general, darker-skinned individuals may present with the pigmentary form.

Papules and nodules are the most common clinical type. Lesions are shiny and pearly, especially at the borders (Figs. 6.8 and 6.9). A central ulceration can be observed, with tumor growth. This neoplasia is located mainly in areas exposed to sunlight: face (cheeks, forehead, palpebral area, and nasolabial fold) and upper chest, but it can occur anywhere on the skin, except on the palms and soles. It may be necessary to confirm diagnosis with a skin biopsy for pathological examination.

Surgical excision is the most effective treatment. Other forms of treatment can be used, such as cryotherapy, electrocoagulation, and photodynamic therapy. Clinical follow-up after treatment is necessary. Indigenous populations present skin phototype III or IV, so the most common presentation is pigmented basal cell carcinoma. This type can be misdiagnosed because it is initially similar to melanocytic nevus.

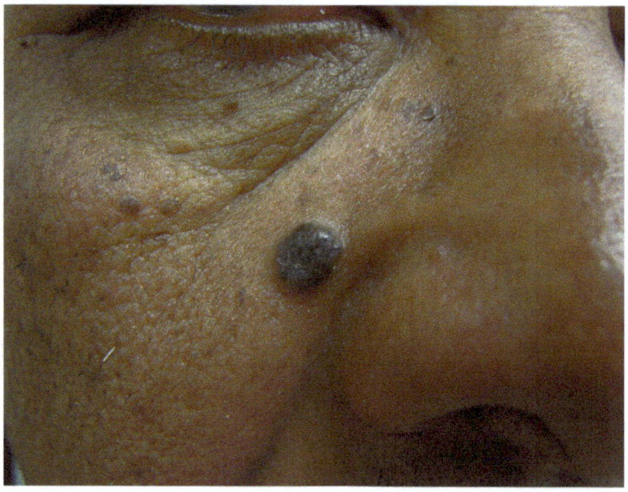

Fig. 6.8 Pigmented basal cell carcinoma: lesion on face

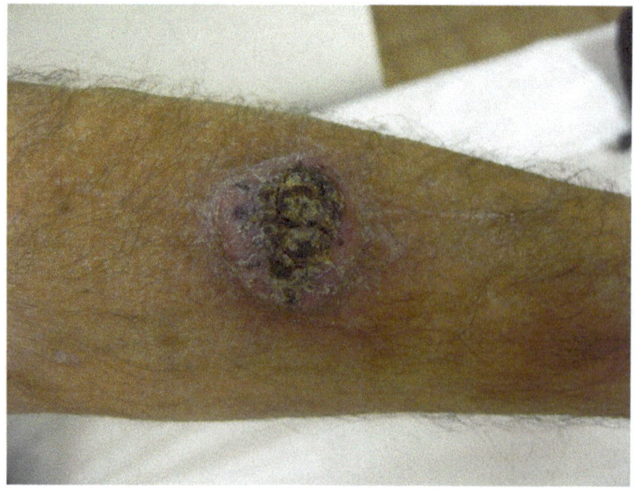

Fig. 6.10 Squamous cell carcinoma: keratotic tumor

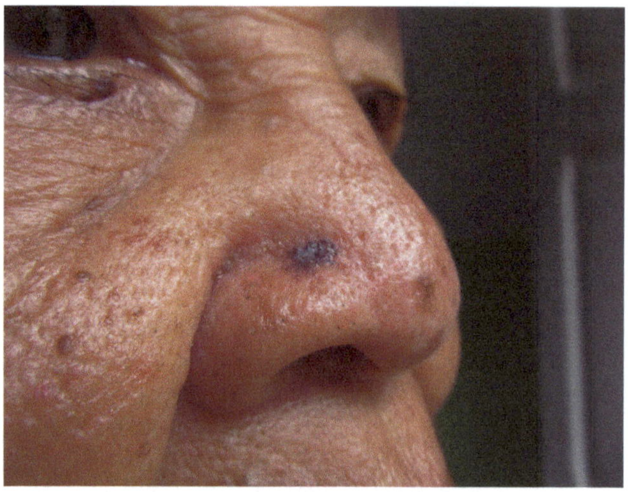

Fig. 6.9 Pigmented basal cell carcinoma: lesion on nose

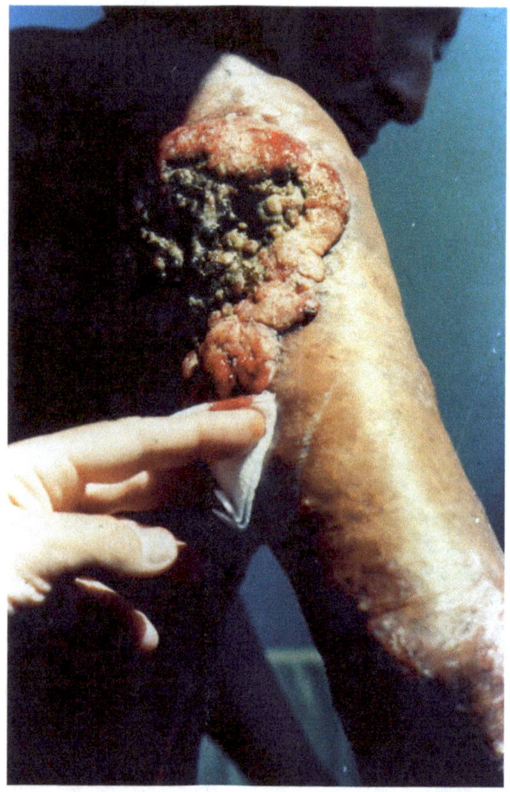

Fig. 6.11 Squamous cell carcinoma: ulcerate tumor

6.6 Squamous Cell Carcinoma

Squamous cell carcinoma is a malignant proliferation of a squamous cell layer. This neoplasia results from genetic predisposition or chronic injuries (ionizing radiation, burns, and chemical exposure). It is also related to low skin phototype, with a history of prolonged sun exposure and infection by oncogenic human papillomavirus.

It occurs in both sexes, with a predilection for men by age 60 years. Elderly individuals are most commonly affected, especially from the sixth decade of life. Certain situations favor its appearance, such as prolonged iatrogenic immunosuppression (solid organ transplant recipients) and genetic diseases such as epidermodysplasia verruciformis, xeroderma pigmentosum, and dystrophic epidermolysis bullosa. Oral squamous cell carcinoma is related to tobacco smoking.

Squamous cell carcinoma is an exophytic erythematous or skin-colored lesion. The size of this neoplasia can vary from that of a papule to that of a tumor. Some lesions are hyperkeratotic, while others can ulcerate with crusts (Figs. 6.10 and 6.11). This carcinoma occurs mainly in areas exposed to the sun, especially on cephalic segments. Frequently preneoplastic lesions, such as actinic keratosis, are observed. It may evolve with metastases in regional lymph nodes.

Diagnosis should be confirmed by histopathological examination. There is an irregular proliferation of eosinophilic keratinocytes with atypical mitoses. In well-differentiated neoplasia there is a large number of horn pearls (eosinophilic parakeratotic keratinization).

In the majority of cases, standard excision is the most effective treatment. Palliative treatments, such as cryotherapy, photodynamic therapy, and curettage with electrodessication, can be used.

In indigenous populations, despite the intense sun exposure without sunscreen photoprotection, the incidence of squamous cell carcinoma is rare.

Suggested Reading

Seborrheic Keratosis

1. Davis SA, Narahari S, Feldman SR, Huang W, Pichardo-Geisinger RO, McMichael AJ. Top dermatologic conditions in patients of color: an analysis of nationally representative data. J Drugs Dermatol. 2012;11(4):466–73.
2. Jackson JM, Alexis A, Berman B, Berson DS, Taylor S, Weiss JS. Current understanding of seborrheic keratosis: prevalence, etiology, clinical presentation, diagnosis, and management. J Drugs Dermatol. 2015;14(10):1119–25.
3. Roh NK, Hahn HJ, Lee YW, Choe YB, Ahn KJ. Clinical and histopathological investigation of seborrheic keratosis. Ann Dermatol. 2016;28(2):152–8.

Melanocytic Nevi

4. Perkins A, Duffy RL. Atypical moles: diagnosis and management. Am Fam Physician. 2015;91(11):762–7.
5. Rabinovitz HS, Barnhill R. Benign melanocytic neoplasms. In: Bolognia JL, Jorizzo JL, Schaffer JV, editors. Dermatology. 3rd ed. Amsterdam: Elsevier Sauders; 2012. p. 1851–83.
6. Roh MR, Eliades P, Gupta S, Tsao H. Genetics of melanocytic nevi. Pigment Cell Melanoma Res. 2015;28(6):661–72.

Nevus Depigmentosus

7. Xu AE, Huang B, Li YW, Wang P, Shen H. Clinical, histopathological and ultrastructural characteristics of naevus depigmentosus. Clin Exp Dermatol. 2008;33:400–5.
8. Kim SK, Kang HY, Lee ES, Kim YC. Clinical and histopathologic characteristics of nevus depigmentosus. J Am Acad Dermatol. 2006;55:423–8.
9. Di Lernia V. Segmental nevus depigmentosus: analysis of 20 patients. Pediatric Dermatol. 1999;16:349–53.

Mucous Cyst

10. De Brito Monteiro BV, Bezerra TM, da Silveira EJ, Nonaka CF, da Costa Miguel MC. Histopathological review of 667 cases of oral mucoceles with emphasis on uncommon histopathological variations. Ann Diagn Pathol. 2016;21:44–6.
11. Delli K, Spijkervet FK, Vissink A. Salivary gland diseases: infections, sialolithiasis and mucoceles. Monogr Oral Sci. 2014;24:135–48.
12. Nallasivam KU, Sudha BR. Oral mucocele: review of literature and a case report. J Pharm Bioallied Sci. 2015;7(suppl 2):S731–3.

Pigmented Basal Cell Carcinoma

13. Brankov N, Prodanovic EM, Hurley MY. Pigmented basal cell carcinoma: increased melanin or increased melanocytes? J Cutan Pathol. 2016;43(12):1139–42. [Epub ahead of print]
14. Hoorens I, Vossaert K, Ongenae K, Brochez L. Is early detection of basal cell carcinoma worthwhile? systematic review based on the who criteria for screening. Br J Dermatol. 2016;174(6):1258–65.
15. Newlands C, Currie R, Memon A, Whitaker S, Wollford T. Non-melanoma skin cancer: United Kingdom National multidisciplinary guidelines. J Laryngol Otol. 2016;130(S2):S125–32.

Squamous Cell Carcinoma

16. Alamgir MM, Jamal Q, Mirza T. Conventional clinical and prognostic variables in 150 oral squamous cell carcinoma cases from the indigenous population of Karachi. Pak J Med Sci. 2016;32(3):672–6.
17. Newlands C, Currie R, Memon A, Whitaker S, Wollford T. Non-melanoma skin cancer: United Kingdom National multidisciplinary guidelines. J Laryngol Otol. 2016;130(S2):S125–32.
18. Soyer HP, Rigel DS, Wurm EMT. Actinic keratosis, basal cell carcinoma and squamous cell carcinoma. In: Bolognia JL, Jorizzo JL, Schaffer JV, editors. Dermatology. 3rd ed. Amsterdam: Elsevier Sauders; 2012. p. 1773–93.

7 Genetic Diseases

7.1 Epidemolysis Bullosa

Epidermolysis bullosa is a group of genetic diseases characterized mainly by the appearance of vesicles and blisters on the skin and mucous membranes, mainly after minor trauma, due to defects in the composition of the cellular ultrastructure of the skin that results in skin fragility. There are forms with localized lesions and extensive lesions (Figs. 7.1, 7.2, 7.3, and 7.4). They are classified as simple, junctional, dystrophic, and Kindler syndrome, depending on the structures and location of the basic defect. These major forms are also subdivided into other forms. In the *simple forms* the formation of blisters is intraepidermal. The clinical presentations are variable and usually do not leave scars, and there is no compromise of the general state. In the *juncional forms*, the formation of blisters occurs at the level of the lucid lamina. The clinical presentations are variable, with some very severe forms that can lead to death by complications. In the *dystrophic forms* the formation of blisters occurs just below the dense lamina of the basement membrane of the epidermis. In addition to the formation of blisters and scars, the nails become affected and the formation of milia is observed.

In some dystrophic forms the condition is more severe and extensive and is associated with weight loss, chronic anemia, synechiae in the hands and feet, and unfavorable prognosis. Squamous cell carcinoma may arise in areas with chronic lesions. Laboratory diagnosis can be made through skin biopsy for anatomopathological examination. To define the level of ultrastructural involvement, other tests are necessary (immunomapping, direct immunofluorescence, immunohistochemistry, or electron microscopy).

For differential diagnosis, the possibility of other genetic or autoimmune bullous dermatoses and bullous impetigo must be evaluated. For therapy, it is important to avoid cutaneous trauma, in addition to the application of dressings, dental treatments, dietary regulation, pain and pruritus treatment, prevention of synechiae, and treatment of complications (e.g., infections).

7.2 Peutz-Jeghers Syndrome

Peutz-Jeghers syndrome is an autosomal dominant inherited disorder. It is characterized by the presence of mucocutaneous lentigines associated with gastrointestinal polyposis. The oral mucosa is the site most suitable for physical examination, contributing to the early detection and diagnosis of the disease. The disease is linked to a higher incidence of colorectal carcinoma This disease is of familial occurrence and emerges after birth or during childhood.

There are multiple brownish to black melanotic and mottled macules (lentigines) present in the oral mucosa. The size of the lesions in general is less than 5 mm (Figs. 7.5 and 7.6). These lesions can also occur on the hands and feet, in the genital region, and around the eyes (Figs. 7.7 and 7.8). Patients present gastrointestinal bleeding and anemia.

Diagnosis is clinical. In a histopathological examination, an acanthosis is observed with increased numbers of melanocytes in the basal layer. There is a heterozygous variant in the STK11 gene that can be detected by a molecular genetic test. The treatment of mucosal lentigines is not necessary. The high incidence of intestinal carcinoma justifies constant investigation by colonoscopy.

7.3 Xeroderma Pigmentosum

Xeroderma pigmentosum is a common group of cutaneous diseases of a genetic character and autosomal recessive inheritance. Eight forms of the disease with variable degrees of change are described.

The basic defect is a lack of DNA repair following exposure to ultraviolet radiation, leading to early skin changes related to sensitivity to sunlight. The result is a greater predisposition to skin cancers and ophthalmological changes. The first symptoms are hyperchromic and freckle spots, telangiectasias, hypochromic spots, and atrophies (Figs. 7.9 and 7.10). Then arise solar keratosis and malignant cutaneous tumors,

M.C. Florian et al., *Dermatological Atlas of Indigenous People*, DOI 10.1007/978-3-319-59446-0_7

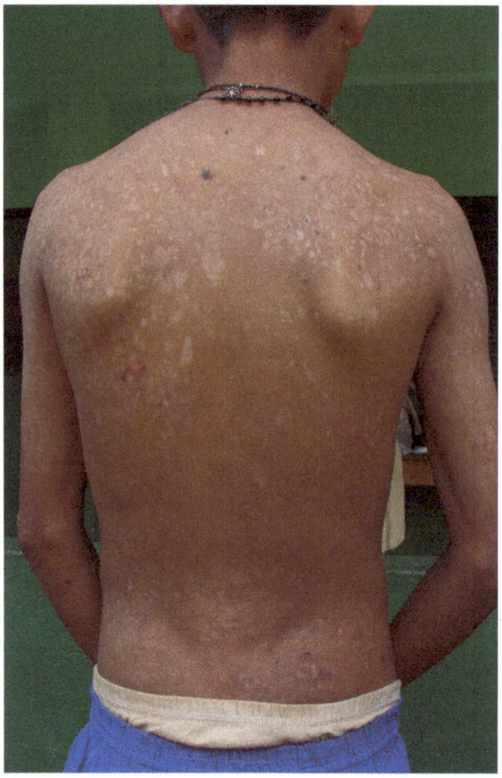

Fig. 7.1 Epidermolysis bullosa: lesions and scars on dorsum

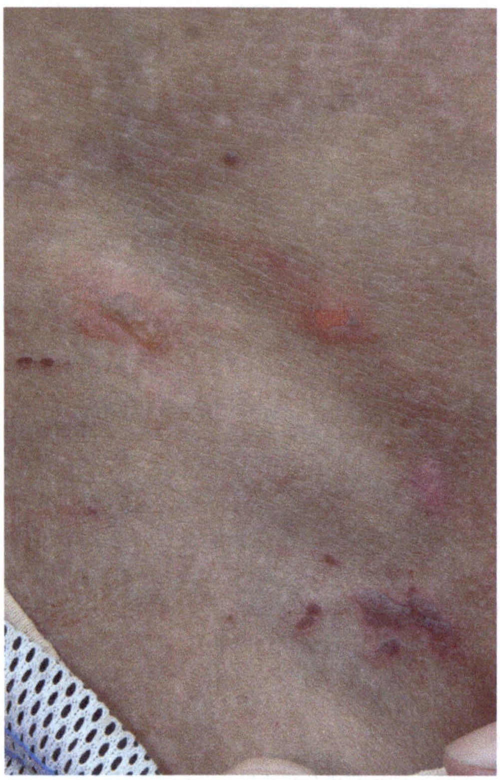

Fig. 7.2 Epidermolysis bullosa: vesicles and blisters

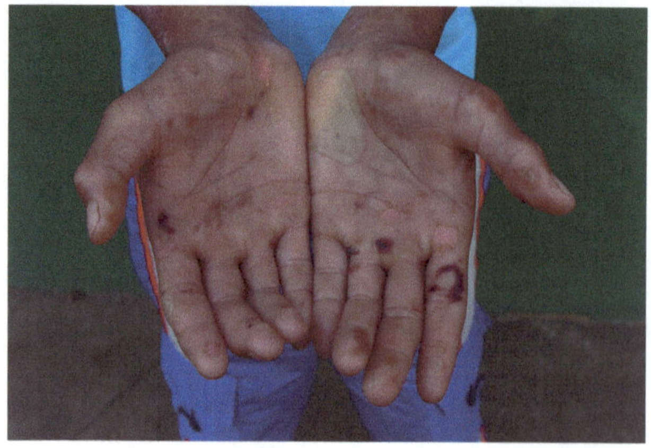

Fig. 7.3 Epidermolysis bullosa: vesicles and blisters on hands

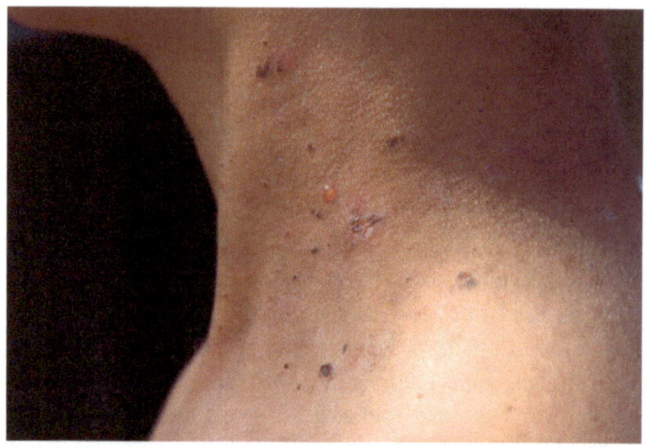

Fig. 7.4 Epidermolysis bullosa: vesicles and blisters on neck

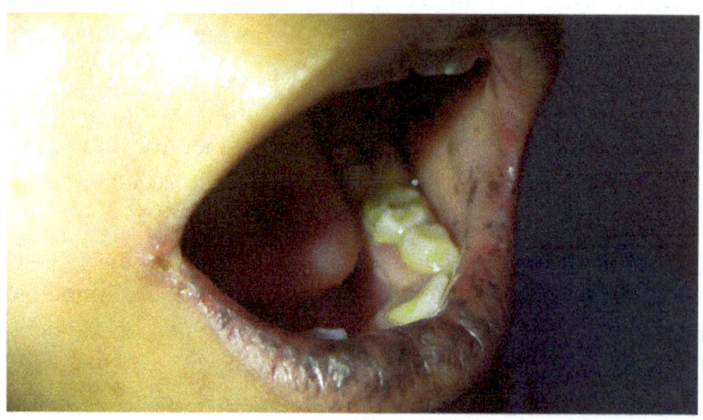

Fig. 7.5 Peutz-Jeghers syndrome: mottled macular lesions on mouth

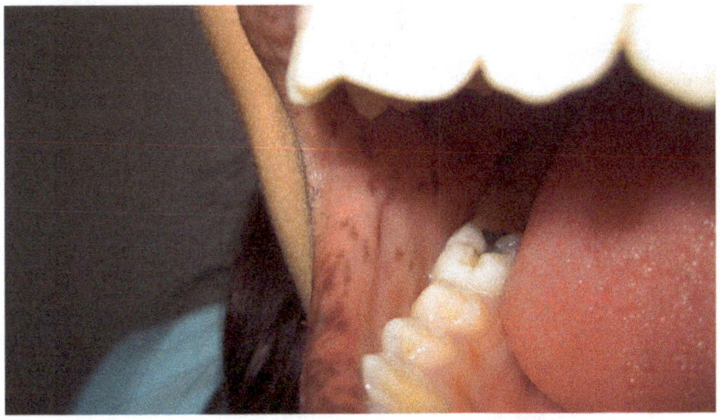

Fig. 7.6 Peutz-Jeghers syndrome: lesions on mouth

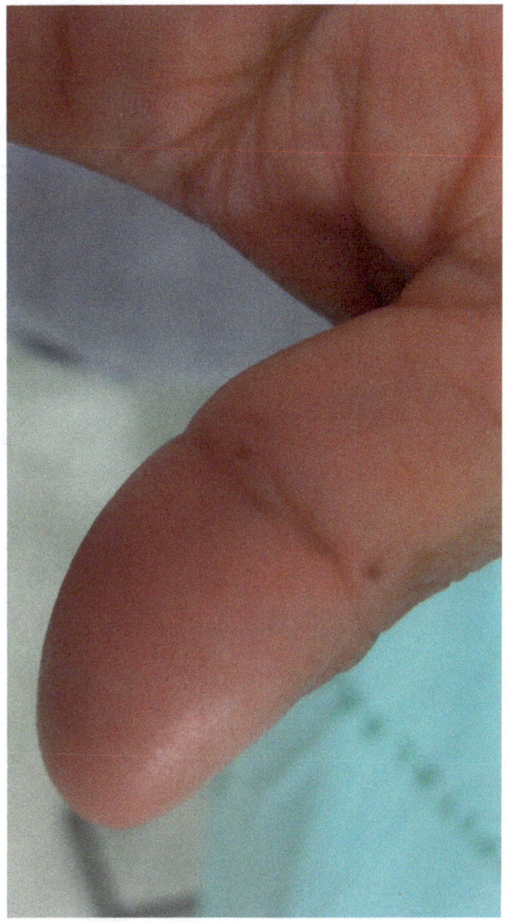

Fig. 7.8 Peutz-Jeghers syndrome: lesions on hand

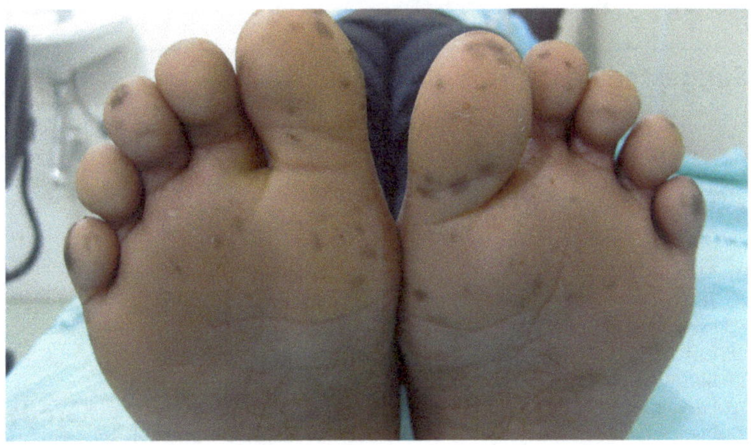

Fig. 7.7 Peutz-Jeghers syndrome: lesions on feet

mainly basal cell carcinomas, squamous cell carcinomas, and melanomas. Other malignant neoplasms (of the skin and other organs) may also occur. Ophthalmological changes related to greater sensitivity to sunlight (e.g., photophobia, conjunctivitis, keratitis, ectropion) may also take place. Neurological disorders occur in about 20% of cases. Severe neurological signs and symptoms may be part of the DeSanctis-Cacchione syndrome.

Diagnosis is clinical. Differential diagnostics must be conducted with other diseases (genetic or not) with photosensitivity characteristics. In general, prognosis is poor because of complications related to skin cancers. Treatment can come in diverse forms of protection against ultraviolet light and sunburn and the treatment of precocious premalignant and malignant lesions. Systemic retinoids (isotretinoin, acitretin) can be used in some cases with variable results.

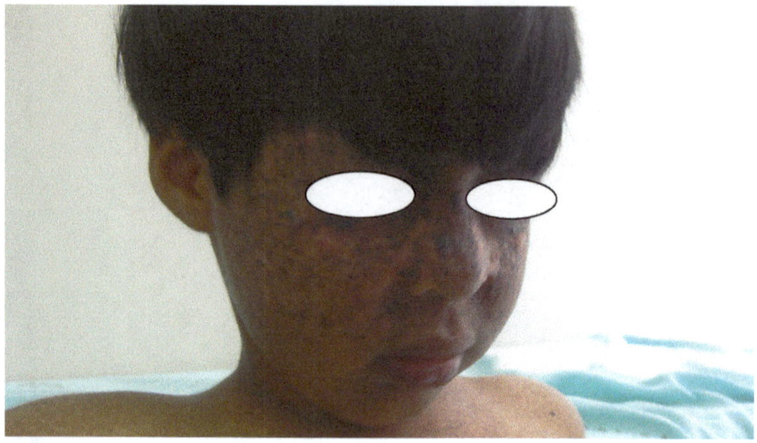

Fig. 7.9 Xeroderma pigmentosum: lesions on face

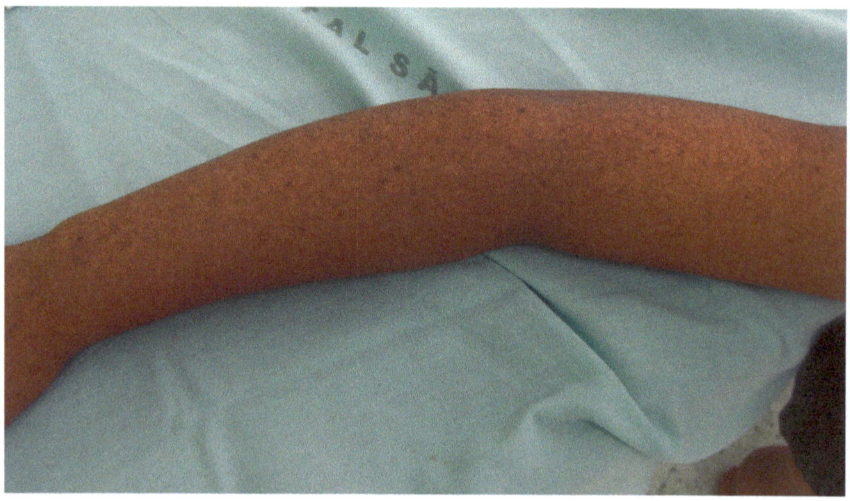

Fig. 7.10 Xeroderma pigmentosum: lesions on upper limb

Suggested Reading

Epidemolysis Bullosa

1. Montaudié H, Chiaverini C, Sbidian E, Charlesworth A, Lacour JP. Inherited epidemolysis bullosa and squamous cell carcinoma: a systematic review of 117 cases. Orphanet J Rare Dis. 2016;11:117.
2. Cohn HI, Teng JM. Advancement in management of epidemolysis bullosa. Curr Opin Pediatr. 2016;28:507–16.
3. Oliveira ZN, Périgo AM, Fukumori LM, Aoki V. Immunological mapping in hereditary epidermolysis bullosa. An Bras Dermatol. 2010;85:856–61.

Peutz-Jeghers Syndrome

4. McGarrity TJ, Amos CI, Baker MJ. Peutz-Jeghers syndrome. In: Pagon RA, Adam MP, Ardinger HH, Wallace SE, Amemiya A, LJH B, Bird TD, Fong CT, Mefford HC, RJH S, Stephens K, editors. GeneReviews [internet]. Seattle: University of Washington; 1993–2016. [updated 2016 Jul 14].
5. Meserve EE, Nucci MR. Peutz-Jeghers syndrome: pathobiology, pathologic manifestations, and suggestions for recommending genetic testing in pathology reports. Surg Pathol Clin. 2016;9(2):243–68.
6. Rabinovitz HS, Barnhill R. Benign melanocytic neoplasms. In: Bolognia JL, Jorizzo JL, Schaffer JV, editors. Dermatology. 3rd ed. Amsterdam: Elsevier Sauders; 2012. p. 1851–83.

Xeroderma Pigmentosum

7. Black JO. Xeroderma Pigmentosum. Head Neck Pathol. 2016; 10:139–44.
8. Lehmann AR, McGibbon D, Stefanini M. Xeroderma pigmentosum. Orphanet J Rare Dis. 6:70; 2011. [Avaliable in: http://www.ojrd.com/content/6/1/70].
9. Viana Fde O, Cavaleiro LH, Carneiro CM, Bittencourt Mde J, Barros RS, Fonseca DM. Do you know this syndrome? Xeroderma pigmentosum (XP). An Bras Dermatol. 2011;86:1029.

8 Miscellaneous

8.1 Hypertrophic Scars and Keloids

Wound healing (after trauma, burning, infections, surgery) is a cutaneous repair process that may result in an atrophic scar or an elevated scar compared to the surrounding skin. The latter is called a *hypertrophic scar*. When the scar is disproportionate to the skin lesion, it is called a *keloid*.

The different forms of scarring are related to individual characteristics and to wound care (e.g., injuries with secondary infections, skin tension in the wound areas). Some ethnic groups may be more prone to hypertrophic scars and keloids, especially those with melanodermic skin.

Clinically, hypertrophic scar and keloid are hard on palpation, with brownish or skin-colored lesions. The hypertrophic scar is an elevated lesion compared to normal skin and do not exceed the original scar (Figs. 8.1 and 8.2). The keloid is a disproportionate scar, attaining a size larger than the scar caused by the original injury (Fig. 8.3).

Diagnosis is clinical. Skin biopsy for histopathological examination should be performed when there are other diagnostic hypotheses. Proposed treatments include topical corticosteroids under occlusion, infiltration of corticosteroids, cryotherapy, or surgery. In case of surgery, it is necessary to take into account the possibility of recurrence of hypertrophic scars or keloids at the site of attempted surgical repair.

Hypertrophic scars and keloids are common among indigenous people. Traumatisms are very common because they are often exposed to the environment and secondary bacterial infection is frequent.

8.2 Lichen Striatus

Lichen striatus is a rare, self-limited dermatosis, with a linear distribution along Blaschko's line. Its etiology is unknown. Children with atopic diathesis are more susceptible to the development of this disease. Genetic and environmental (virus) factors are likely responsible for lichen striatus onset.

More than 50% of cases occur in children under 15 years old, but onset can occur at any age. It affects women more often than men.

Clinical features are characterized by small bright, whitish, or purple papules, which increase rapidly, converging in papular plates in a linear arrangement, restricted to a few centimeters in size or covering the whole affected limb (Figs. 8.4, 8.5, and 8.6). Nails may also be affected, with permanent dystrophy (Fig. 8.7). Mainly the disease affects the upper and lower limbs. Generally, the disease is asymptomatic, but some patients report pruritus.

Diagnosis is based on the clinical presentation. If necessary, a skin biopsy that shows a lichenoid inflammation with dyskeratotic keratinocytes may be performed.

In most cases treatment is not necessary. If there is itching, topical or intralesional corticosteroid infiltration can be prescribed. Cream and moisturizing oils should be used. The lesions regress spontaneously within 3 months to 1 year.

Interestingly, we observed 6 cases of lichen striatus in Xingu Indigenous Park, a large number considering about 400 patients were examined. The disease is common in children, with no gender preference. One patient presented digital involvement with nail dystrophy. There was no predominant ethnic group.

8.3 Endemic Pemphigus Foliaceus ("Wild Fire")

Endemic pemphigus foliaceus, or *fogo selvagem*, is a bullous disease characterized by the production of IgG antibody against anti-desmoglein 1. This antigen is located on the surface of keratinocytes in the most superficial portion of the epidermis (granular layer). This specific antibody causes the rupture of desmosomes (acantholysis), and a cleavage occurs, resulting in this bullous disease. The endemic pemphigus foliaceus is a regional disease in Brazil and is probably caused by environmental factors. Currently, the triggering of "fogo selvagem seems to be linked to the black flies (*Simulium* spp.), common in these areas.

M.C. Florian et al., *Dermatological Atlas of Indigenous People*, DOI 10.1007/978-3-319-59446-0_8

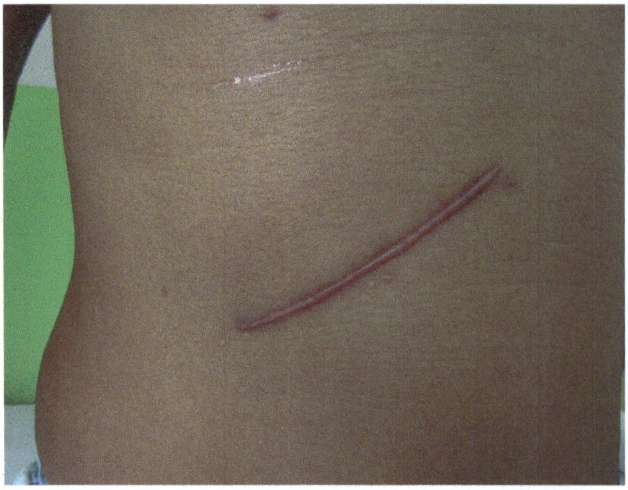

Fig. 8.1 Hypertrophic scars: Linear lesion on abdomen

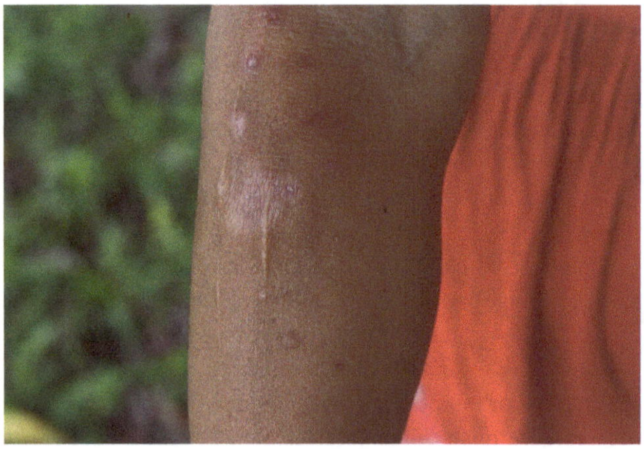

Fig. 8.2 Hypertrophic scars: Multiple linear lesion on upper limb

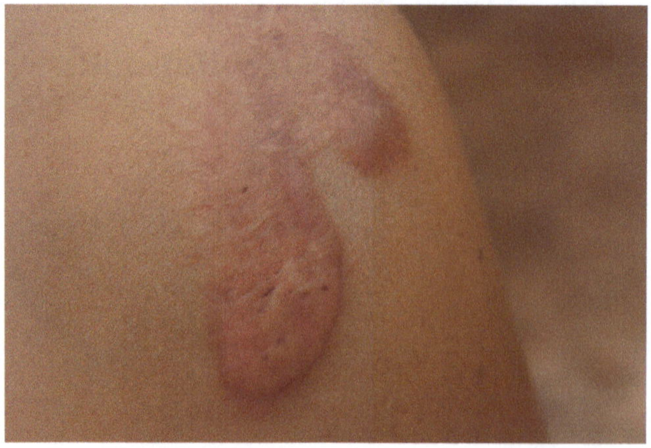

Fig. 8.3 Keloids: Enlarged lesions on shoulder

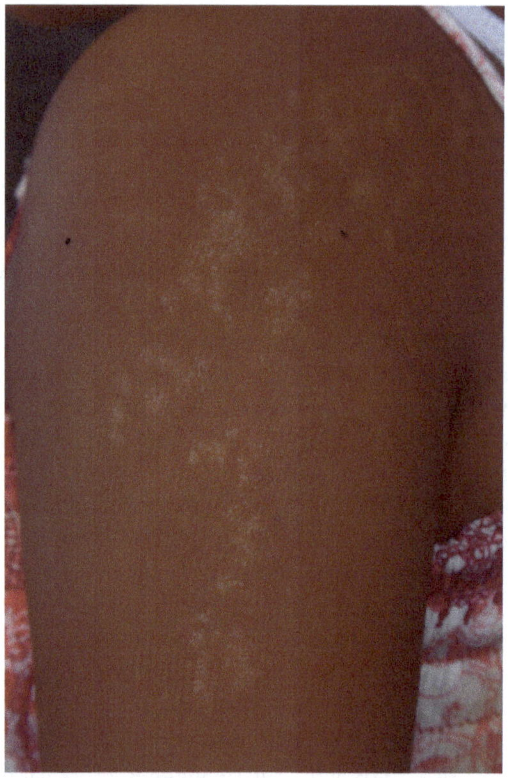

Fig. 8.4 Lichen striatus: Linear distribution whitish papules on arm

Pemphigus foliaceus, unlike pemphigus vulgaris, occurs in children and young adults. There is no gender preference. People exposed to this environment, especially in rural areas, can develop fogo selvagem.

Superficial cleavage results in fragile vesicules and bullous lesions that easily rupture (Fig. 8.8). Therefore, clinical examination will show mainly erosions, crusts, and desquamation (Fig. 8.9). The lesion distribution is similar to that observed in seborrheic dermatitis (Fig. 8.10). A few patients develop exfoliative erythrodermic forms. There is no mucosal involvement, as observed in pemphigus vulgaris.

The histopathology shows acantholysis in the upper epidermis (granular layer), forming blisters. Direct immunofluorescence in skin fragments shows IgG deposition on the keratinocyte cell surface. Indirect immunofluorescence examines sera and detects circulating IgG antibodies against epithelial cell.

Treatment is similar to that for pemphigus vulgaris. The majority of cases are treated with oral corticosteroids or dapsone. Chronic and mild lesions are treated with potent topical corticosteroids. Immunosuppressive drugs are recommended in aggressive forms.

Frequently, indigenous populations are exposed to the disease in rural areas and near water. There are several reports of fogo selvagem among indigenous people.

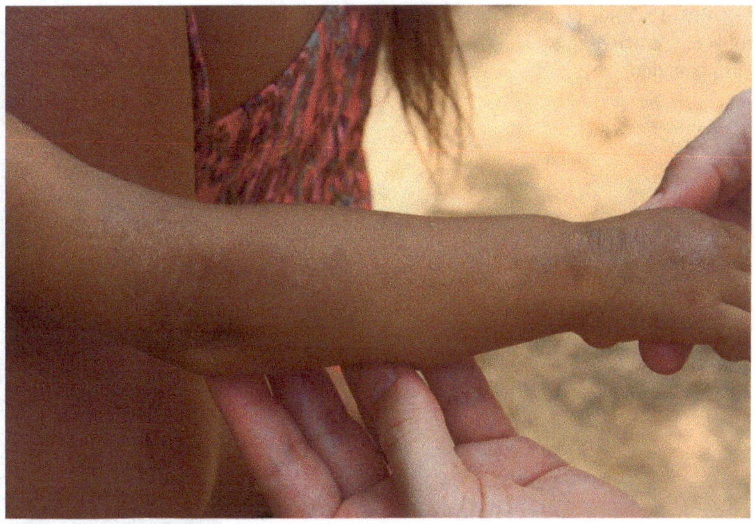

Fig. 8.5 Lichen striatus: lesions on upper limb

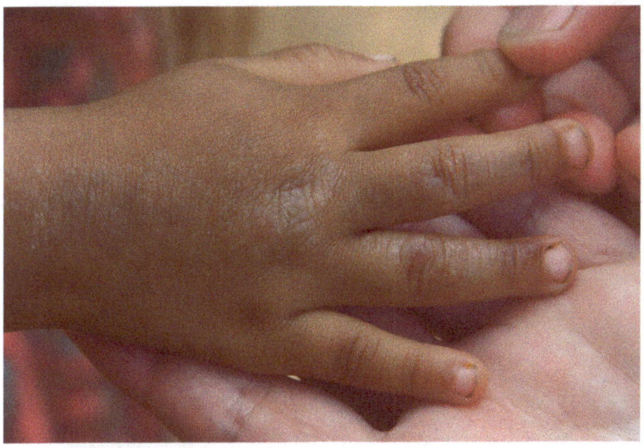

Fig. 8.6 Lichen striatus: lesions on upper limb

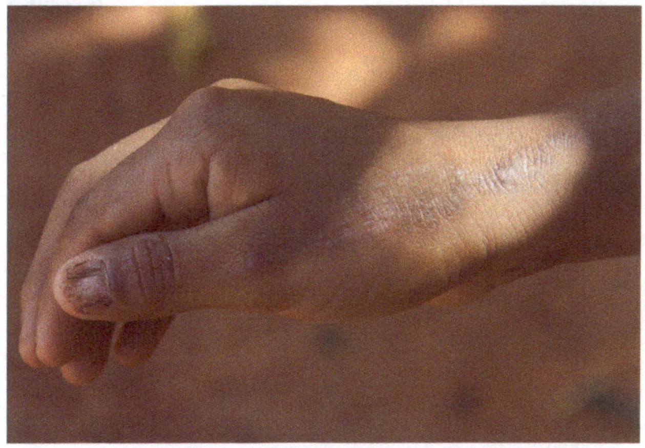

Fig. 8.7 Lichen striatus: nail dystrophy

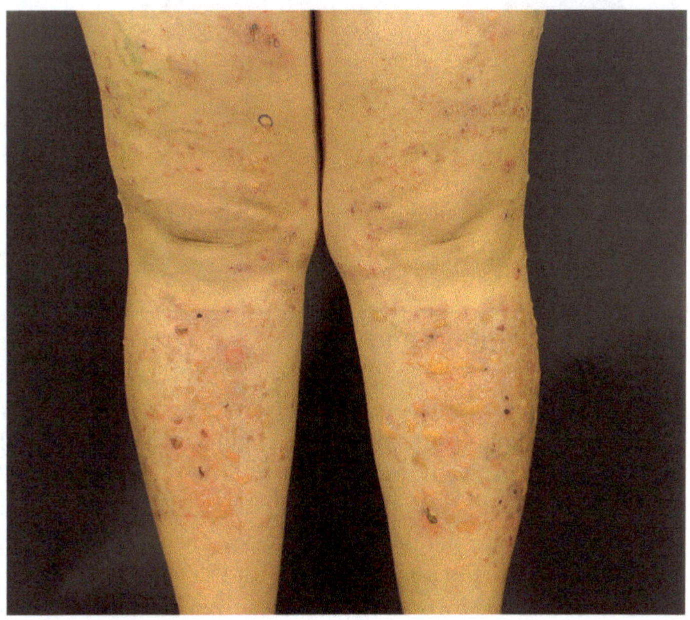

Fig. 8.8 Endemic pemphigus foliaceus: vesicles and blisters

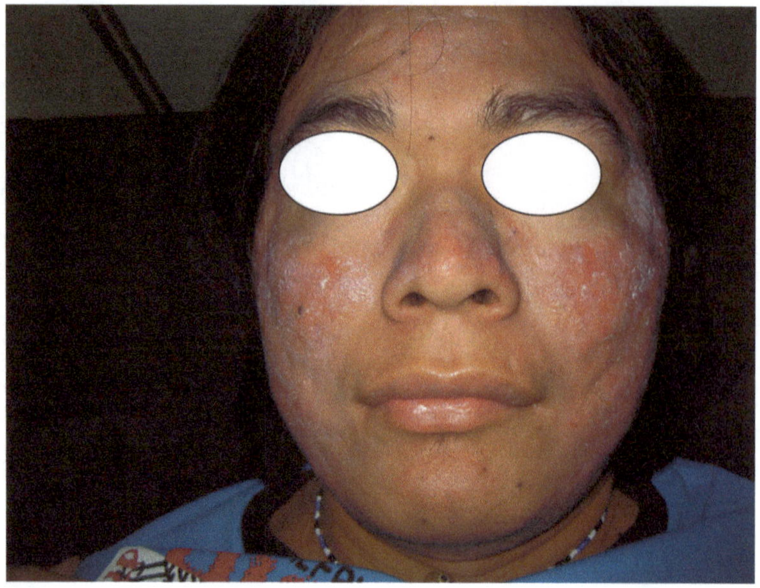

Fig. 8.9 Endemic pemphigus foliaceus: erosions and desquamation

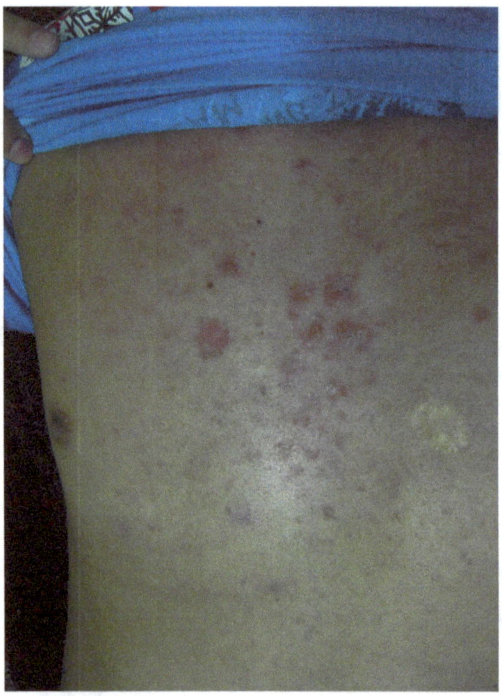

Fig. 8.10 Endemic pemphigus foliaceus: lesions on seborrheic areas

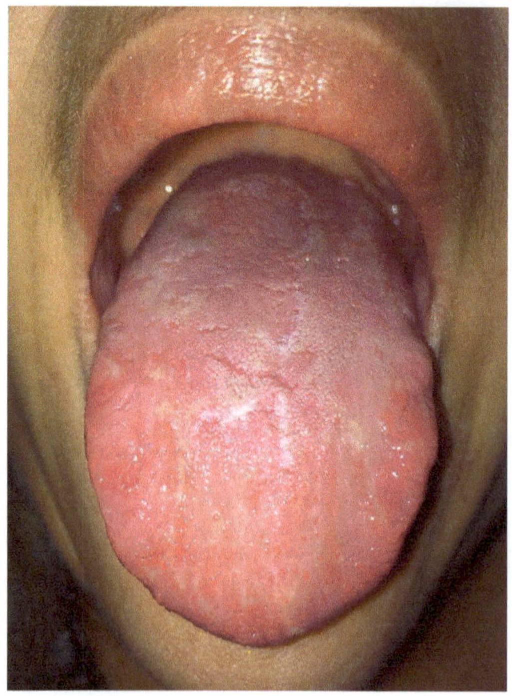

Fig. 8.11 Geographic tongue (Photo by Dalva Regina Neto Pimentel)

8.4 Geographic Tongue

Geographic tongue is also called benign migratory glossitis. It is a recurring, benign disease of familial occurrence. Its etiology is unknown. Previously it had been related to psoriasis because the pathology shows psoriasiform mucositis. Its prevalence in the general population is about 1–3%. It preferentially affects children but can start at any age. It is manifested by outbreaks characterized by areas of papillae loss, forming a map of "geographical" aspect on the tongue (Fig. 8.11). Patients report that these areas seem to migrate (migratory characteristic). Some patients complain of burning at the site of injury, especially with hot or spicy foods.

Diagnosis is made by clinical examination. The histopathology shows a psoriasiform mucositis, but biopsy is not required for diagnosis; it is usually not necessary. For local burning or difficulty eating, topical corticosteroids in orabase presentation or mouthwashes are useful. As an asymptomatic disease, it was probably not the main complaint during the consultation in Xingu Indigenous Park.

8.5 Fox-Fordyce Disease

Fox-Fordyce disease is a rare disease that involves obstruction of the apocrine sweat gland. It is a pruritic disease, with no infectious agent involved. The actual etiology is not known. It occurs almost exclusively in females, usually between 14 and 35 years of age, with perimenstrual flare. The epidemiologic data support a hormonal role.

Individualized skin-colored papules are located at the exits of hair folllicules associated with itching (Fig. 8.12). The common sites are the axillae, the pubic area, mammary areola, and the inguinal–crural region. Pregnancy and the use of contraceptives decrease the symptoms. The presence of a keratin plug in the follicular infundibulum could help the histopathological diagnosis. Recently, a perifollicular foam cell infiltrate was described in the disease.

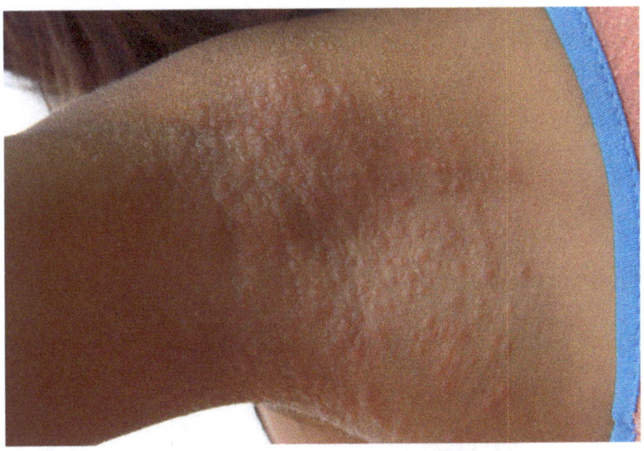

Fig. 8.12 Fox-Fordyce disease: lesions on armpit

Some of the treatment options include oral contraceptives, topical retinoic acid, and topical corticosteroids. In some cases, surgery may be necessary. Cases of very precocious onset are known.

8.6 Hydroa Vacciniforme

Hydroa vacciniforme is a vesicular and bullous disease on photoexposed areas that tends to appear during the warm and sunny months. This disease occurs worldwide. Nowadays, it is established that it is caused by human herpesvirus 4 (HHV4) or Epstein-Barr virus (EBV). The disease occurs mainly in children and usually disappears after puberty.

The lesions develop symmetrically on photoexposed areas, such as the face (nose, cheeks, ears), but they can affect the extremities (dorsal aspect of hands). Pruritic papules develop into vesicules and bullous lesions. Several days later, hemorrhagic crusts evolve into atrophic scars.

Skin biopsy could help with diagnosis, and direct immunofluorescence is nonspecific, but it is necessary to differentiate it from other diseases (porphyria, polymorphous light eruption, lupus erythematosus).

The majority of patients are refractory to treatment. Measures should be taken to protect against sunlight (e.g., hats with protection, sunscreen), and as well as beta carotene and antimalarials should be given. Figures 8.13 and 8.14 show an Indian child with typical manifestation of hydroa vacciniforme. No disease related to EBV infection had previously been observed in Xingu Indigenous Park. Prolonged sun exposure in this geographic region could facilitate the development of this rare disease.

Fig. 8.13 Hydroa vacciniforme: lesions on face

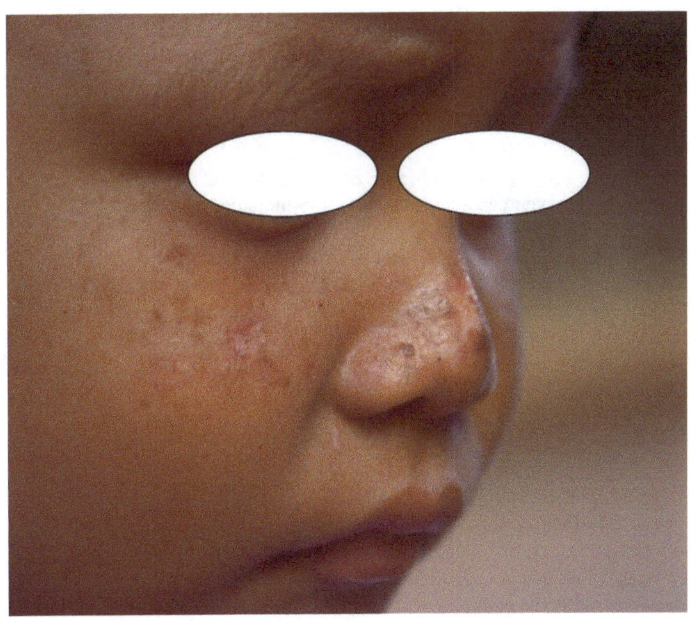

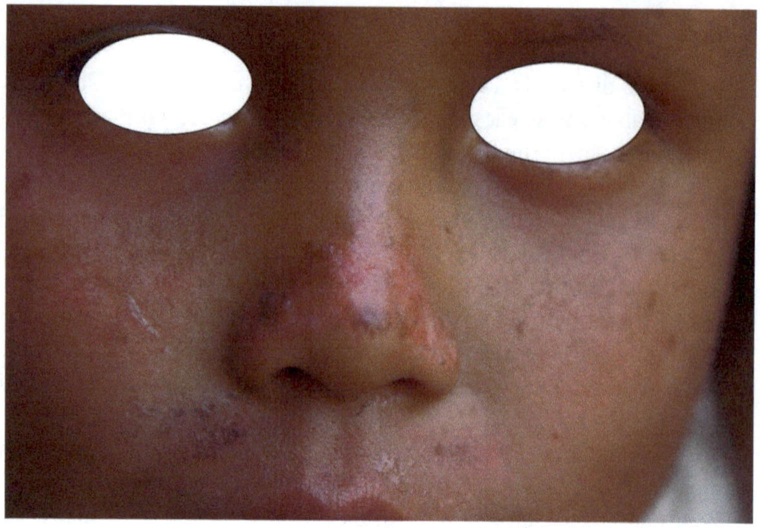

Fig. 8.14 Hydroa vacciniforme: lesions on face

Suggested Reading

Hypertrophic Scars and Keloids

1. Andrews JP, Marttla J, Macarak E, Rosenbloom J, Uitto J. Keloids: the paradigm of skin fibrosis – pathomechanisms and treatment. Matrix Biol. 2016;51:37–46.
2. Berman B, Maderal A, Raphael B. Keloids and hypertrophic scars: pathophysiology, classification, and treatment. Dermatol Surg. 2016;24:S3–S18. [Epub ahead of print]
3. Trace AP, Enos CW, Mantel A, Harvey VM. Keloids and hypertrophic scars: a spectrum of clinical challenges. Am J Clin Dermatol. 2016;17(3):201–23.

Lichen Striatus

4. Johnson M, Walker D, Galloway W, Gardner JM, Shalin SC. Interface dermatitis along Blaschko's lines. J Cutan Pathol. 2014;41(12):950–4.
5. Kim M, Jung HY, Eun YS, Cho BK, Park HJ. Nail lichen striatus: report of seven cases and review of the literature. Int J Dermatol. 2015;54(11):1255–60.
6. Müller CS, Schmaltz R, Vogt T, Pföhler C. Lichen striatus and blaschkitis: reappraisal of the concept of blaschkolinear dermatoses. Br J Dermatol. 2011;164(2):257–62.

Endemic Pemphigus Foliaceus ("Wild Fire")

7. Alvarez RR, Campbell I, Friedman H, Bertoli ML, Gama GB, Diaz LA. Dermatosis among the Xavánte from the pimentel barbosa Indian reservation, mato grosso, (Brazil). Article in portuguese. Cad Saude Publica. 1991;7(4):581–4.
8. Aoki V, Rivitti EA, Diaz LA. Coorperative group on fogo selvagem research. Update on fogo selvagem, an endemic form of pemphigus foliaceus. J Dermatol. 2015;42(1):18–26.
9. Friedman H, Campbell I, Rocha-Alvarez R, et al. Endemic pemphigus foliaceus in native Americans from Brazil. J Am Acad Dermatol. 1995;32(6):949–56.

Geographic Tongue

10. Assimakopoulos D, Patrikakos G, Fotika C, Elisaf M. Benign migratory glossitis or geographic tongue: an enigmatic oral lesion. Am J Med. 2002;113(9):751–5.
11. Mangold AR, Torgerson RR, Rogers RS 3rd. Diseases of the tongue. Clin Dermatol. 2016;34(4):458–69.
12. Picciani BL, Domingos TA, Teixeira-Souza T, et al. Geographic tongue and psoriasis: clinical, histopathological, immunohistochemical and genetic correlation – a literature review. An Bras Dermatol. 2016;91(4):410–21.

Fox-Fordyce Disease

13. Brau Javier CN, Morales A, Sanchez JL. Histopathology attributes of fox-fordyce disease. Int J Dermatol. 2012;51(11):1313–8.
14. Yost J, Robinson M, Meehan SA. Fox-fordyce disease. Dermatol Online J. 2012;18(12):28.
15. Kao PH, Hsu CK, Lee JY. Clinicopathological study of fox-Fordyce disease. J Dermatol. 2009;36(9):485–90.

Hydroa Vacciniforme

16. Di Lernia V, Manouri Y. Epstein-Barr virus and skin manifestations in childhood. Int J Dermatol. 2013;52(10):1177–84.
17. Millard TP, Hawk JL. Photosensitivity disorders: cause, effect and management. Am J Clin Dermatol. 2002;3(4):239–46.
18. Nitiyarom R, Wongpraparut C. Hydroa vacciniforme and solar urticaria. Dermatol Clin. 2014;32(3):345–53.

Index

M.C. Florian et al., *Dermatological Atlas of Indigenous People*, DOI 10.1007/978-3-319-59446-0